Studies in Economic Ethics and Philosophy

Springer-Verlag Berlin Heidelberg GmbH

Studies in Economic Ethics and Philosophy

P. Koslowski (Ed.)
Ethics in Economics, Business, and Economic Policy
(out of print)
192 pages. 1992

P. Koslowski · Y. Shionoya (Eds.)
The Good and the Economical
Ethical Choices in Economics and Management
212 pages. 1993

H. De Geer (Ed.)
Business Ethics in Progress?
124 pages. 1994

P. Koslowski (Ed.)
The Theory of Ethical Economy in the Historical School
345 pages. 1995

A. Argandoña (Ed.)
The Ethical Dimension of Financial Institutions and Markets
264 pages. 1995

G. K. Becker (Ed.)
Ethics in Business and Society.
Chinese and Western Perspectives
232 pages. 1996

Peter Koslowski

Ethics of Capitalism and Critique of Sociobiology

Two Essays with a Comment by James M. Buchanan

Springer

Professor Dr. Peter Koslowski
Forschungsinstitut für Philosophie Hannover
Gerberstraße 26
D-30169 Hannover, Germany

Part Two of this book has been translated from the German by David Ambuel.

This book is an English edition of the following two books. Springer-Verlag would like to thank the original publisher J.C.B. Mohr (Paul Siebeck) Tübingen for granting permission to publish them in the English language.

Peter Koslowski, Ethik des Kapitalismus
© 1982 by J.C.B. Mohr (Paul Siebeck) Tübingen

Peter Koslowski, Evolution und Gesellschaft
Eine Auseinandersetzung mit der Soziobiologie
© 1984 by J.C.B. Mohr (Paul Siebeck) Tübingen

Cataloguing-in-Publication Data applied for

Die Deutsche Bibliothek - CIP-Einheitsaufnahme

Koslowski, Peter:
Ethics of capitalism and critique of sociobiology : two essays with a comment by James M. Buchanan / Peter Koslowski. [Part two of this book has been transl. by David Ambuel].

(Studies in economic ethics and philosophy)
Enth.: Ethics of capitalism [Einheitssacht.: Ethik des Kapitalismus <engl.>]. Critique of sociobiology [Einheitssacht.: Evolution und Gesellschaft <engl.>]. - Enth. ausserdem: The morality of capitalism / by James M. Buchanan
ISBN 978-3-642-08246-7 ISBN 978-3-662-03311-1 (eBook)
DOI 10.1007/978-3-662-03311-1
NE: Buchanan, James M.: The morality of capitalism

ISBN 978-3-642-08246-7

Originally published by Springer-Verlag Berlin Heidelberg New York in 1996
Softcover reprint of the hardcover 1st edition 1996

SPIN 10535235 42/2202-5 4 3 2 1 0 - Printed on acid-free paper

Preface

This book publishes the English translation of texts that appeared first in German in two separate booklets. Part One, *The Ethics of Capitalism* including the comment by JAMES M. BUCHANAN, has been published in German under the title *Ethik des Kapitalismus*, Tübingen (J.C.B. Mohr [Paul Siebeck]) 1982, 5th edition 1995, in the series „Walter Eucken Institut, Vorträge und Aufsätze", vol. 87. Part Two, *Evolution and Society. A Critique of Sociobiology*, appeared first in German under the title *Evolution und Gesellschaft. Eine Auseinandersetzung mit der Soziobiologie*, Tübingen (J.C.B. Mohr [Paul Siebeck]) 1984, 2nd edition 1989, in the same series, vol. 98.

Part One of the book has been translated by the author, Part Two by DAVID AMBUEL.

I should like to thank Liberty Fund Inc., Indianapolis, USA, for its support of the translation of Part Two, Georg Siebeck of J.C.B. Mohr (Paul Siebeck) Publishers, Tübingen, for the permission to publish the English translation of the two essays and to my co-workers at the Forschungsinstitut für Philosophie Hannover - The Hannover Institute of Philosophical Research, Hannover, Germany, for their support in editing this volume.

Hannover and Herdecke, December 1995 Peter Koslowski

Contents

CONTENTS

CONTENTS

Part One

The Ethics of Capitalism

Preface to the Second German Edition

The necessity of republishing this book as early as one and a half years after the first publication gives the author the opportunity to add - inspired by the discussion of the book - some explanatory remarks on the text that has been revised and enlarged in some parts. The book at hand has been developed from a lecture given at the conference „Philosophical and Economic Foundations of Capitalism" held at Freiburg im Breisgau, Germany, February 1st to 4th, 1981. The conference was sponsored by Liberty Fund Inc., Indianapolis, USA, a foundation aiming at encouraging studies about the fundamental principles of a society of free and self-responsible humans. The lectures of this conference together with a short version of Part I of this book have been published by Svetozar Pejovich under the title *Philosophical and Economic Foundations of Capitalism* (Lexington Mass.: D.C. Heath, 1983).

The subject originally given to me was „The Morality of Capitalism", a subject that one would not so easily set himself within the German social philosophical discussion for two reasons:

On the one hand, the title „The Morality of capitalism" sounds to German ears more or less like an oxymoron, that is a composition of contradictory terms. We do not use the concept of capitalism with the neutrality of the Americans but mostly pejoratively, connecting it rarely with the concept of morality.

On the other hand, it is the scope and generality of the question whether the capitalistic order is moral-ethically justifiable that rather deters from dealing with it within the usual division of labour between the disciplines of philosophy and economics, a division of labour that is, however, changing only more recently.

Thus, the small booklet at hand cannot be expected to deal with the subject in an exhaustive manner. It rather has the character of a programmatic writing that has to be followed by its elaboration in a comprehensive work about the social philosophy and political economy of a free society.[1] It is to be hoped, however, that the book raises the relevant questions and makes clear that the problem of the right economic order has to be investigated in a

---

1 I have developed my approach to economic ethics and to a cultural and philosophical theory of the economic order in KOSLOWSKI (1988), KOSLOWSKI (1991), and KOSLOWSKI (1994).

cooperation between social philosophy and political economy and that, in turn, the problems of social philosophy cannot be solved without the insights of political economy.

With respect to this interest in a synthesis of social philosophy, political economy, economic ethics and theory of the economic order the author considers himself to be close to the Freiburg School of political economy, especially to the works of Walter Eucken and of Friedrich August von Hayek whose work, in a wider sense, belongs also to this school. He is grateful to the Walter-Eucken-Institut for publishing the second edition of this work, as the first edition, in the series of the institute.

I have retained the term capitalism in the title and in the formulation of the question although several scholars have given me the advice to substitute it by the less emotionally charged concept of „market economy". I have not followed this proposal because the term market economy with regard to my question is analytically not sharp enough. Unlike „capitalism" the term „market economy" does not describe a theory of society or a concept of the totality of the economic order, but the more limited aspect of the market as an instrument of coordination that can be effective in noncapitalistic societies as well. Capitalism, on the other hand, represents an entire societal order being characterized by the market economy, private property, and economic individualism (individual profit- and utility maximiziation as the goal of economic action). Thus, for grasping a total societal order and its ideal model in a comprehensive manner the term capitalism is more suitable than the term market economy.

The investigation will, however, show that this model of capitalism as a social theory is incomplete and needs to be supplemented by a comprehensive social philosophy and a social ethic including the state. Since the time of Alfred Müller-Armack, the theory, transforming capitalism and supplementing it by a political and social framework that compensates its deficiencies as a social order by political structures and institutions of social policy, has been named by the term „social market economy". If a theory of the social market economy is meant to be more than a slogan, its foundations must lie in the synthesis of the fundamental principles of capitalism such as the subjectivism and freedom of utility and profit maximization with an individual and social ethic based on natural right. The theory of the economic order (*Ordnungstheorie*) of the social market economy has to be sought in a synthesis of economic liberalism and the older tradition of natural right.

A reviewer of this book from the University of Chicago, Michael Gillespie, wrote in *Ethics* (Vol 94, 1983) that the synthesis intended here is questioning positions of freedom and subjectivity that modernity can not surrender. In response to this critique it has to be emphasized that this book is dedicated to the preservation of the heritage of modern subjectivity, but it is also led by the conviction that, in the face of the alarming loss of meaning and environmental problems in western societies, the securing of freedom in the modern society and economy is no longer possible without this synthesis of the older tradition of natural right and of modern economic liberalism. Whether this synthesis intended of the older natural right tradition and of economic liberalism is „at the heart of modernity“ seems to be as well undecided as irrelevant for those who interprete modernity per se not as an argument but primarily as a description of a place in time, namely the place of the historically latest. After all, the approach of a reconciliation of the tradition of freedom and of natural right can, under conditions of post-modernity, perfectly claim to be modern.

The question whether the synthesis intended in this book or the total liberation of subjectivity in the economy is the proper position of modernity is philosophically not very rewarding and theoretically undecidable. We cannot know which position will outlive the other. The tendencies in economic theory to re-emphasize the importance of the humanities within the formation of economic theory, as we can find them in recent approaches of the theory of the firm and of business ethics, confirm the claim that economic positivism or pure neoclassical economics are, despite their undisputable contribution, no longer sufficient for the understanding of our economic realities.

University of Munich, March 1984

Preface to the Third German Edition

The need for a third edition of this book, again one and a half year after the last edition, demonstrates that the approach presented here found the interest of the economic as well as of the philosophical profession. Possibly, this approach corresponds to a state of consciousness of the above mentioned postmodern. The text is, beside the correction of some minor printer‘s errors, taken from the second, revised edition.

University of Witten/Herdecke, January 1986

Preface to the Fourth German Edition

The developments, if not the revolution, of the year 1989/90 in Europe have changed the whole concept of „capitalism“. Until now we have taken the pejorative meaning of capitalism for granted. This understanding is no longer valid. The laws of development claimed by Marxism that predicted the evolution of developed capitalism to socialism and communism changed into the laws of development from developed or „really existent“ socialism to capitalism. Communism turned out to be a very painful detour to capitalism. This development is no reason for triumph, but for a re-thinking of the foundations of capitalism. The question of the ethics of capitalism gains new urgency.

Forschungsinstitut für Philosophie Hannover - Hannover Institute of Philosophical Research, Hannover, Germany, November 1990

Preface to the Fifth German Edition

The book in hand is forthcoming in the Japanese and Chinese language as well. To the degree that capitalism is about to become a world phenomenon, the investigation of its ethical and cultural foundations turns out to be a question of high priority for humankind. The dialogue with the non-western cultures and societies about the question whether capitalism is the only possible economic order in the world or whether it is or must be modified by the different cultures of the world, is one of today's central tasks and questions. Will there be, as some Japanese voices claim, a „Christian capitalism“, a „Confucian capitalism“, an „Islamic capitalism"etc.?

Forschungsinstitut für Philosophie Hannover - Hannover Institute of Philosophical Research, February 1st, 1995

Introduction

Who would not be bright enough to see much in his surroundings which is, indeed, not as it should be?

HEGEL,
Enzyklopädie der philosophischen Wissenschaften (1830), § 6

The world is always only a little short of salvation.

CARL STERNHEIM, 1913,
Aus dem bürgerlichen Heldenleben

Within the scope of studies on the philosophical and economic foundations of capitalism, the moral inquiry into the ethics and morality of capitalism is certainly most delicate and ambiguous. It must find a path between uncritical apology and presumptuous moralism, between precipitate acceptance of the status quo and abstract imperatives.

Capitalism as an economic order is distinguished by three structural characteristics, namely, private disposal of means of production, market and price mechanisms as means of coordination, and profit and utility maximization as the basic motivation in economic action. The thesis of this book is that capitalism is inseparably bound with the history of freedom and subjectivity and constitutes a necessary component of a free society but that a theory of capitalism as a societal form which considers the capitalist economic order to be the whole of society falls short of societal reality. All attempts to base society exclusively upon these three structural characteristics fall to the reductionism objection as raised already by neoliberals such as W. Röpke and A. Rüstow. Tendencies of the contemporary positivist economic theory[1] to make the capitalist economy and its paradigm a universal and conclusive theory of human action and society, indeed, beyond sociobiology to even make it a theory of all life forms, present an interesting theory imperialism of economics but are ultimately economic reductions. These theories cannot ground the conditions for preservation of capitalism; instead they endanger them.

---

1 BECKER (1974); HIRSHLEIFER (1978).

In believing that they can dispose with ethics and with the posing of value questions, contemporary positivist economists overlook that as a society of free individuals, capitalism places enormous moral demands on the individuals and requires a moral attitude that the economy alone cannot produce. Against such attempts it is necessary to recall that economics originated from moral philosophy and that its father Adam Smith wrote two treatises: *The Wealth of Nations* and *The Theory of Moral Sentiments.* A social philosophy of capitalism must have the same breadth of perspective as Adam Smith. It must guard against committing the economic fallacy of believing that an economically efficient system already makes for a good or moral society and that the economy is the whole of society.[2]

Part I of this book will first give an overview of the ethical, moral-theological, and metaphysical changes that accompanied the advent of capitalism, that is, a sketch of its *normative genesis.* Then the questions of whether capitalism requires ethics or is at all compatible with ethics and whether capitalism can be justified as a social system will be considered.

Chapter I

What Does Morality of an Economic System Mean?

„Morality of an economic system" looks like a contradiction in terms, a *contradictio in adiecto*. The economy as the system for the provision of material goods ought to meet economic, not moral norms. For the sake of the efficient production and distribution the economy even ought to be kept free of well intended, but hindering, moral reasoning. These postulations, however, already imply an „ought", a counterfactual, normative element. The economy itself is not an „ethical neuter", since it is not only determined by economic laws, but also by humans whose will and choice are guided by a combination

2 Compare from the German neoliberal tradition RÖPKE (1949) and RÜSTOW (1945). Recently BELL (1976). The crisis of capitalism is not caused, as Bell shows, by economic inefficiency or the superiority of other systems but by the crisis of cultural and ethical integration in the Western world.

of manifold expectations, norms, mental attitudes, and, last but not least, by moral conceptions. Human action ensues from highly complex conceptions, economic reasoning forming just one part of them.

The way of looking at the economy as at an autonomous social system is of modern origin and has become apparent only since Mandeville.[3] In traditional, premodern societies economic action is inseparably connected with religious, family, and political action. It is not until the rise of capitalism that the economy becomes an autonomous sphere of society. Is this process a process of liberation, a process that can be justified in moral terms, or is it a process of dehumanization and alienation? Marx interpreted this process as both liberating and alienating. According to him capitalism destroys all „feudal, patriarchal, idyllic relationships", but on the other hand capitalism, and capitalism alone, has proved, „what human activity is able to bring about and has achieved miracles totally different from Egyptian pyramids, Roman aqueducts, and Gothic cathedrals. [....] All that is corporative and fixed is vaporizing, all that is sacred is desecrated, and men finally are forced to look at their situation, their mutual relationships with sober eyes." (translated from the German original)[4] So if capitalism is not only an economic, but also a social phenomenon, the question of its morality has to be trailed beyond the borderlines of individual ethics and can be met only by a holistic view.

The question of the morality of capitalism cannot contribute an additional moral aspect to the economic, sociological, and political aspects of the topic of capitalism; rather, it must be understood as the integration and moral evaluation of the totality of arguments. Morality is not one aspect among others, but a way to appreciate the perspectives and arguments of the sciences, to order and evaluate them, and to render them meaningful for human action.[5] The question of the morality of capitalism cannot be: „Is capitalism moral?" Rather, the question must be: „Is capitalism justifiable under the conditions of human nature and the scarcity of resources?"

A principle of the moral theology and natural right of the Baroque period stated that moral obligation arises from the nature of the object.[6] The moral-

3 DUMONT (1977).

4 MARX (1848), p. 464 f.

5 This paper's concept of morality follows the reinterpretation of natural right which has been suggested by ROBERT SPAEMANN (1980, pp. 39-40) most recently.

6 Compare LUIS DE MOLINA: *De justitia et jure*, Moguntiae (1602), Vol. 1 disput.4 n.2: „obligatio juris naturalis oritur a natura objecti" (the obligation

ity of capitalism can be justified only by the nature of the object, that is, the function of the economy and the possibilities it offers for human self-realization. Morality amounts to appropriateness to the matter and cannot consist in the abstract opposition of a moral ought as against economic arguments. As the distinction between descriptive and prescriptive propositions was never fully maintained and an economic natural right of efficiency always prevailed in economics, the question of the morality of capitalism need not fear the naturalistic fallacy because this objection has in some sense itself proved a fallacy. The moral inquiry is not opposed to economic theory but must take up the latter and ask whether all aspects of reality are done justice.

To the understandable objection that this claim to a totality of perspectives in ethics is very extensive, one must answer that people raise the question of the justification of their actions and of the system in this universality and not as an inquiry into single aspects of their existence. In addition, one lives in a totality of social conditions and is determined by them. One would want neither to live in a just society where there is nothing to buy nor in an efficient, rich society that employs its resources for morally reprehensible purposes. In the inquiry into the ethics of capitalism and into the totality of its characteristics, therefore, scientific precision must not be paid for by the renunciation of the entirety of possible aspects. At the same time it is evident that a social order can never be justified once and for all because the number and importance of the viewpoints by which it must be evaluated constantly change with time.

Chapter II

The Development Toward Capitalism

1. The Liberation of the Structural Features of Private Property, of Maximization of Profit and Utility, and of Coordination by Markets

The attempt to suitably define *capitalism* causes difficulties since the notion of capitalism is to a great degree laden with negative connotations. An *ism*

---

of natural law arises from the nature of the object); n.3: „Obligatio oritur a natura rei" (the obligation arises from the nature of the matter).

generally signifies an exaggeration of an otherwise legitimate notion.[7] Consequently, the concept of capitalism, as Röpke justly remarks, „has been made responsible for all suffering, evil and injustice in the world, all of which were beforehand put into the concept of capitalism"[8]. Several examples from the 111 different definitions of capitalism that Passow[9] lists may clarify this. Capitalism is there designated as the spirit of haggling and usury, as the commercialization of society, as mammonism, as the bastard of feudal landed property and free market, and as the victory of plebeians and parvenus.

In contrast to these abuses, occidental capitalism is defined here by three essential structural characteristics which are not limited to capitalism but which bear a particularly distinguished form in it:

(1) *private property—of means of production as well;*

(2) *profit and utility maximization as economic purposes*; and

(3) *coordination of economic activities by markets and prices.*[10]

All three structural characteristics can be found in all societies, traditional and modern, but they assume special distinction in capitalist societies. People always have private property at their disposal—at least on their arms and hands. They always pursue the goal of maximum profit besides other goals, and the relative prices for goods and services reflect always, besides other determinants, supply and demand ratios and, thereby, relative scarcity.

According to Joachim Röpke[11], three exchange mechanisms are present in *all* societies: exchange via markets, exchange by coercion or the threat of coercion, and forms of reciprocal, socially determined exchange (gifts, relations of exchange in families, and so forth). In principle, the types of exchange do not alter, but in modern times and especially in pure capitalism the relative predominance of market, redistribution, and reciprocity relationships shifts toward the market, whereas reciprocal relations that are determined by social and cultural values prevail in traditional societies. Röpke's thesis is confirmed by Brentano's thesis that market exchange started from the trade with foreigners at the boundaries at a state and in overseas trade.[12] It is also supported by the research of Sahlins who shows that patterns of exchange in tra-

7 This might be more so the case in the German than in the English language.

8 W. RÖPKE (1949), p 45.

9 PASSOW (1927) p. 12.

10 For this definition see KROMPHARDT (1980) p. 38.

11 J. RÖPKE (1970), p. 15.

12 BRENTANO (1902), p. 18.

ditional societies depend on degrees of kinship to the extent that market principles are applicable only to intertribal trade; the cultural overlapping with and social standardization of economic acts of exchange increases in proportion to the intensity of kinship.[13]

The degree of the embeddedness of the market in social and cultural norms and the extent of social and cultural restrictions under which the price system works are higher in traditional than in modern societies. Modern capitalism is distinguished not by its structural characteristics but by the extent of their *disembeddedness.*[14] This is not only true for the price system but also for profit maximization and property. Maximization of payoff[15] is a fundamental distinction of human rationality, but the *freeing* of the structure of commercial motivation from religious and cultural contexts is the essential feature of modernity which asserts itself in Europe during the Renaissance and mercantilism. It is, therefore, not only capitalistic, but this feature marks the beginning of the economic age altogether. It is not *pleonexia*, that is, nonsatiety, which signifies the advent of capitalism but the moral neutralization of profit maximization as a respectable motive and driving force of the economy.

The process of moral neutralization and social disembeddedness extends to property rights as well. Property is an anthropological constant given by individual corporeality and the individual's disposition over his corporeal instruments. The conceptual pleonasm *private property* which Hegel introduced into the philosophical discussion shows a new shade of property rights. Property has become private, its social restrictions reduced, its private rights extended. As J. G. Fichte realized, the extension of material property rights as free rights of the owner is necessarily accompanied by the abolition of property rights on persons (bondage, serfdom), which means a reembeddedness of property under the moral norm of personal freedom.[16]

---

13 SAHLINS (1965), p. 97.

14 The degree of the embeddedness of the market in social and cultural norms is higher in traditional societies as POLANYI (1971) demonstrates. Compare DUMONT (1977), SAHLINS (1965), and TAWNEY (1927). Cf. for the economic history of capitalism that is not treated here SOMBART (1902) and (1912) and HARTWELL (1983).

15 See below Part Two, chapter III.2, for a discussion of the generality of the maximization principle and its extension from maximization in human decision-making and under rational foresight to maximization as a principle of natural evolution.

16 J. G. FICHTE, *Der geschlossene Handelsstaat* (1800). Compare MAX WEBER (1972), p. 61.

In summing up, the development toward capitalism can be described as the process of the autonomization of the economy (Edgar Salin)[17], or as the dissolving of economic relations from the social and cultural norms in favour of a stronger self-regulation by the economy, a stronger influence of economic laws on distribution and the social structure, and a stronger orientation of the status system, the social appreciation, and the stratification of society according to the economic performance evaluated by the market results. It is this emancipation of the economy that evokes anticapitalist critique and procapitalist applause. The question of the morality of capitalism is, therefore, essentially an inquiry into the legitimacy of the moral and cultural neutralization of the three structural characteristics.

The process in which the economy becomes autonomous is a manifestation of the development of the European spirit toward individualization, subjectivization, and rationalization. Thus, it is bound up with the history of freedom in western culture. Individualization means the freeing of man from the constraints of inherited social status and from determination by social and religious norms.This process involves a differentiation between the individual and society, which the individual perceives simultaneously as liberation and alienation. It is an indication of the development toward subjectivity. The self, the I, not the group, becomes responsible for his or her actions and social position. Western individualism finds its basis in the developmental tendency from status to contract and from ascription to achievement.

The origins of western individualism lie, as German idealistic philosophy repeatedly emphasized, in Christianity. The high order of subjectivity and selfhood is religiously founded in the Christian conception of each man as an image of God, the incarnation of God in the person of one man, the personal character of God, the individual judgment after death, and bodily (individual) salvation. In assertions such as „Christ set us free, to be free men“ (Gal. 5:1), the pathos of subjectivity in Christianity comes into force; it means first of all the moral autonomy of the subject against the law but cannot be restricted to this form of autonomy.[18]

The freedom of the moral subject requires alignment of the economic process with the individual plans of producers and consumers. The indivisibility

---

17 For SALIN the term *economy* should only be applied to the autonomous economy (1967, p. 2).

18 Trail-blazing work was done by the philosophy of history and religion of HEGEL and by TROELTSCH (1923). Cf. for more recent works JOHNSON (1979). The different sources of Japanese capitalism are explored in KOLM (1985).

of freedom demands freedom of commerce and economic value-subjectivism. The contract between consenting partners as the basis for employment and exchange relationships, and the setting of prices according to the individual's subjective evaluation and willingness to pay, reflect the development toward autonomy and subjectivity. Employment relations by assigned status and administered prices correspond, on the other hand, to objectified visions of the economy.

The process of individualization and subjectivization is closely allied with that of rationalization, as Max Weber described it.[19] The freeing of individual pursuit of goals and the unrestrained coordination of these goals via the market lead to a form of rationalization which no longer establishes socially obligatory, universal goals and then attempts to socially realize them with minimal costs. This form of rationality attempts with a minimum of expenditure to realize arbitrary goals which are not socially and culturally fixed. Society and economy no longer act value rationally, that is, concerning material values to be reached with the minimum expenditure, values that are integrated by way of social values. Rather it acts goal rationally (Max Weber's *zweckrational*), that is, rationally concerning all individual goals, which are integrated into a single market by way of price signals.

Value-rational acting does not need to be acting under duress, if the values are generally accepted, but it does not derive its criteria of rationality from subjective purposes, but from objectified purposes. The formal rationality of the market corresponds to pure catallactics, whereas value rationality corresponds to a finalized economy, to the household model of the economy.

There is a certain tension between autonomy and value-rationality because the individual regards his or her own judgment as the standard of value and will find it difficult to accept social values. Within Christian religion the tension between individualization, autonomy and value-rationality first appeared in the Reformation.[20] In Protestantism and in Catholicism, the tension between individualization, autonomy and value-rationality has led to divergent conceptions of authority and freedom, of church community and individual, to different forms of ecclesiastic and thereby social integration, and to a different conception of science.

Max Weber has shown that there is a connection between the doctrine of justification and economic pursuit of success, particularly in Calvinism, and that there is a connection between capital accumulation and ascetic Protes-

---

19 Compare MAX WEBER (1972), pp. 44-45, with KOSLOWSKI (1976), p. 69.

20 MAX WEBER (1979).

tantism in the development of capitalism, but he has overemphasized the confessional differences within Christianity, so that too little attention has been paid to the features generally common to Christianity, namely individualization, autonomization, and rationalization.

The development of occidental culture and society toward capitalism and toward autonomization is laden with a dialectic of freedom, and this, as is often forgotton in economic positivism, is a capacity for both good and evil. The question for social philosophy therefore, is, to what extent the moral and social neutrality of private property, of the maximization of profits, and of the price mechanism can be justified.

2. Freeing of the Economy from Moral Theology and Ethics

The autonomization of the economy is a moral problem, as is the autonomization of morality. Historically, the question of whether the economy should be separated from social norms has been a problem of moral theology and has first appeared in the moral theological discussion about freedom of contract and the ban on interest. There are two prerequisites for freedom of contract: the internalization of external effects in business contracts and the responsibility of the contracting parties. Assertions that *both* conditions are fulfilled have often proved to be unfounded, and therefore the justification of freedom of contract within moral theology, too, could be carried out only through harsh intellectual arguments.

a) Freedom of Contract and Just Price

According to the Aristotelian and the medieval doctrine a contract has to come about not only in a free, but also in a just way. According to the doctrine of just price, not only the goods bartered by contract ought to correspond in value, but also the advantages of the contracting parties ought to be balanced. In judging a contract, the social status and the neediness of the contracting parties has to be taken into account . This quite restrictive concept of *iustitia distributiva* was not applied to all contractual relationships in business though. For many of them only the weaker standards of *iustitia commutativa* were applied. The transition to the capitalistic paradigm here as elsewhere is not an immediate one, but elements of capitalism are already to be found in traditional societies.

In contrast to that, it is characteristic for the capitalistic paradigm that the question of justice as to a contract is not asked any more, but every actual contract is regarded as a justified one. Hobbes expressly rejects the concept of distributive justice. He contrasts a definitely nominalistic concept of value to the „realistic" one of the Aristotelian tradition.

The agreed price of a good is also its value. The value is determined by the agreement of the contracting parties alone. „The value of all things contracted for, is measured by the appetite of the contractors: and therefore the just value, is that which they be contented to give."[21] In contrast to that Thomas Aquinas had held that buyer and seller betray each other, if „the seller sells a good at a price exceeding its value, or the buyer buys at a price lower than the value of the good".[22] This theory does not make sense to Hobbes: „As if it were Injustice to sell dearer than we buy; or to give more to a man than he merits."[23] According to Hobbes the contractual agreement alone determines the value of a thing, and the contractual price of a thing is also its value.

In Hobbes's view moral reflection about contracts is pointless. A contract either complies with the law or it does not. If it complies with it, it is just, too. Justness of a contracting party does not consist in anything else but in performing a contract, it does not consist in taking into account the status of the partner or the „ontological" value of the contractual object. „To speak properly, Commutative Justice, is the Justice of a Contractor; that is, a Performance of Covenant".[24]

Every contract all contracting parties have consented to without being forced to is a just contract. Justness is not defined by a correspondence between price and ontological value, but by the consent of all parties, by their wilful decisions. Consenting to a contract implies mutually advantageous barter and thus justness of the contract. According to Dumont[25], the concept of mutually advantageous barter did not arise until mercantilism, whereas beforehand barter had been considered nearly exclusively as a zero-sum-game. This change constitutes a decisive point and indicates the rise of the paradigm of the capitalistic economy.

---

21 TH. HOBBES: *Leviathan* (1651), ed. C. B. Macpherson, London (Penguin) 1968, chap. 15, p. 208. See also MACPHERSON (1962), p. 78 ff.

22 THOMAS AQUINAS, *Summa theologiae* (abbr. *s.th.*), 2-2, q.77 a.1,1.

23 HOBBES, *Leviathan*, chap. 15, p. 208.

24 *Ibid.* See also KOSLOWSKI (1982), p. 166 ff.

25 See DUMONT (1977), p. 35.

The capitalist understanding of contractual freedom tends to be blind to certain side effects of a moral character. One could characterize these side effects as „moral externalities", external effects of and to the moral qualities of both parties.

For example, the seller can take account of the buyer's financial stress in the setting of price and practice price discrimination in favor of the buyer. On the other hand, in certain situations indifference of price may be a moral demand of fairness itself.

In a restricted market with heterogeneous goods the distribution of producer and consumer rents is not fixed. Moral externalities are possible. On the other hand, the capitalist paradigm requires the large market with homogeneous goods and the absence of price discrimination, that is, a unitary market price in order to compensate for the moral externalities. According to Adam Smith, whether or not one can obtain meat should not depend on the friendliness or moral quality of the butcher. The old paradigm of justice of contract left the answer to this question to the personal moral judgment: depending on the situation, price discrimination or unitary market price can be morally obligatory.[26]

b) Externalities as a Moral Problem: Probabilism

Side effects are not only a problem for the economy but also a central problem for morality altogether. Freedom of commerce and of contract could only be introduced after it had been proven that external effects, by and large, are internalized so that the pursuit of self-interest - Mandeville's private vices - could bring about positive external effects - public benefits.[27] The question of the permissibility of an action or a contract when there is the possibility of external effects brought about a lengthy moral discussion in the sixteenth and seventeenth centuries, the probabilism dispute. It concerned the question of which authority should be followed when one desires to conclude a contract of questionable (moral) permissibility. May possible side effects be accepted or not?

The ethical discussion in the probabilism dispute distinguished the following possibilities: When the party does not conclude a dubious contract, she or he follows the morality and theologically more certain opinion (*opinio*

26 Of course, not to discriminate in prices can be a moral obligation as well.

27 HIRSCHMAN (1977) describes the Christian forerunners of Mandeville who show that the ideological transition to capitalism was gradual. See for a discussion of the ethics of the just price KOSLOWSKI (1988).

tutior). When the party concludes it despite his or her doubts, she or he follows the less certain opinion, the *opinio minus tuta*. Whoever concludes a contract for whose permissibility stronger reasons can be given than those opposed follows the probable opinion (*opinio probabilior*). It is remarkable that probabilism—also named laxism—allows an action when weaker reasons, but nevertheless some reasons, speak for its permissibility.[28]

This view first appeared thoroughly stated in the *Treaties on the Contracts of Merchants* at the end of the fifteenth century.[29] The establishment of laxism by the Jesuits in the seventeenth to eighteenth centuries coincided with the rise of capitalism. The weakening of the rigoristic demands of *tutiorism* concerning the permissibility or nonpermissibility of side effects goes hand in hand with the establishment of freedom of contract. The strict moral conception of just contract corresponds to a stationary, conservative economy, and the subjectivistic conception of morality, that is, the probabilistic morality corresponds to a dynamic economy and to the system of free entrepreneurship.

R.H. Coase[30] has shown that certain discoveries and innovations are economically feasible only when an exclusion or at least a restriction of the liability of producers for side effects is established. It follows that far-reaching technical and social innovations can hardly be introduced if tutiorism prevails, whereas probabilistic criteria may pay too little attention to social costs and externalities. Capitalism, as a system of contractual freedom and technical innovation, historically required the weakening of rigoristic morality and the toleration of external effects. Until now, the problem of externalities has not found a satisfying solution as the example of environmental pollution shows.[31] The moral and theological changes toward probabilism are part of a process of subjectivization and individualization described above.

---

28 See DÖLLINGER/REUSCH (1889), p. 4, and SPAEMANN (1977).

29 JOANNES NIDER, *Tractatus de contractibus mercatorum*, Paris circa 1495 (Munich: Bayerische Staatsbibliothek, Sign. Nr. 4 1341), English translation: JOHANNES NIDER: *On the Contracts of Merchants*, translated by Charles H. Reeves, edited by Ronald B. Shuman, Norman (University of Oklahoma Press) 1966.
The article „Probabilisme" in: *Dictionnaire de Theologie Catholique* (Paris 1936), 13,1, co. 446, calls Nider one of the earliest forerunners of probabilism. Compare KOSLOWSKI (1984).

30 COASE (1960)

31 Cf. WEGEHENKEL (1981). Cf. for the problem of side effects or externalities also KOSLOWSKI (1984).

The problem of interest belongs as well to this moral context and could be regarded as a special case of the freedom of contract. Because of its meaning it must, however, be treated as a problem of its own.

c) The Prohibition of Charging Interest and Freedom of Contract

The decontrol of charging interest sheds light on the process of freeing the economy from social norms. According to the Scholastic theory of interest, interest is morally forbidden because it is a surreptitious price for the sale of a common good that belongs to all, namely time.[32] The interest-usurers want to regard time as the gift in return for which they receive the income that lies in the addition to their loan formed by the amount of interest.

According to Aegidius de Lessinia, their intention can be derived from the fact that they increase or decrease their claim to interest payments according to the time for which a loan has been granted: the longer is the time span of the loan the higher the higher is its price in terms of interest. Time, however, is a common good which does not specially belong to a single person, but is given by God equally to all humankind. By asking a price for the time as the good he gave away the usurer betrays his neighbour who possesses the sold time in the same way as the usurer himself, and he betrays God for whose free gift he asks a price.

Aegidius de Lessinia has in a nutshell a time preference theory of interest. Time preference, however, should not be a tradeable good, but should be spared out from the kind of goods for which freedom of contract is applicable. The contracting partners have no freedom of contract in matters of interest since time is a common good.

Interest as the price for loans should not be formed by market forces but should be bound by the objective norms of equality and justice. The Scholasticists support hereby a static form of the economy and society when they allow interest-free consumption loans but forbid investment loans against the payment of interest. It is indeed difficult to decide, whether the prohibition of interest stabilizes and promotes a stationary economy, or whether it exists as a moral command *because* of the static character of the economy for which it

---

32 AEGIDIUS DE LESSINIA: *De usuris in communi* (1276/85), cap. 4, in: THOMAS AQUINAS: *Opera Omnia* (ed. Fretté), Paris 1875, Vol. 28, p. 581. This work was attributed for a long time to Thomas Aquinas, thus by BÖHM-BAWERK (1884), p. 18. - J. Nider writes as well that time is a good that one is not allowed to sell.

was given and in which private loans are in fact only granted for bridging periods of true neediness and the prohibition to charge interest therefore protects the weaker contracting party.

The discussion about the prohibition of interest has been led in the moral-theological debate between the Dominican and the Jesuit order. It is closely linked to the theological controversy whether men does the good out of grace and predestination or out of freedom. In this debate the Jesuits defended human freedom against the thesis of predestination theory that human action is determined by divine predestination and the freedom to take interest against the prohibition of interest. The Dominicans, however, argued for the depedence of man on grace, for the predominance of grace over freedom, and for the prohibition of interest.[33]

The anthropological-theological view that man is bound to the grace of God coincides with the boundedness of the medieval economy. If one adds the probabilism in ethics as the specific Jesuit position in moral theology or ethics to this picture a process to greater freedom and free-setting of subjectivity identical in the moral theology and ethics and in economics becomes visible.[34]

---

33 Cf. KNOLL (1967), p. 37/38. Compare however also HÖFFNER (1955) who attributes to the entire Scholastic period a greater openness towards a dynamic economy but extends Scholasticism until modern authors like Luis Molina.

34 The relationship between the boundedness and unboundedness of the morals and the economics of a society, of interest and grace, can be seen at the example of the probabilist moral theologian Johannes Caramuel y Lobkowitz (died 1682). According to BÖHM-BAWERK (1884), p. 18 note 1, Caramuel y Lobkowitz was the first to advocate a time preference theory of interest that „present money is more valuable than future money“ and thereby a subjectivist theory of interest. In moral theory he was one of the most radical laxists - celebrated by Antonio Diana as *princeps laxistarum*. According to DÖLLINGER/REUSCH (1889), p. 30, he left so much room to subjectivity and freedom of the moral judgement that the acting person could make by taking recourse to *opiniones probabiles* that what had been sin in former times is now not sin anymore.

3. The Metaphysics of Capitalism: *katallaxia* versus *oikonomia*

The question of which economic system is best is not only logically related to the question of whether this is the best of all possible worlds. The answers to both questions always depend on the metaphysics which one advocates and are in a certain sense themselves metaphysical answers. The term metaphysics is used here to signify the set of the last and fundamental, empirically nonrefutable principles of the constitution of the objects of cognition and of cosmology that lie at the base of our world perception. Each culture has its own social metaphysics in which its last and general principles of „Weltanschauung" are included. This social metaphysics exerts a decisive influence on the conception of the economic system.[35]

The basic principle of social metaphysics is the correspondence of social and cosmic structure. In western societies this connection is not easily recognized because as pluralistic societies they bring forth a pluralistic metaphysics or what William James called a „pluralistic universe".

a) Central Direction of the Economy: *oikonomia*

The concept of economics has its roots in the Greek *oikonomia*, the household and the administration of the house. *Oikonomia* and *oikonomika* originally meant the knowledge of the master of the house (*pater familias*) as to the direction of the house and to the rule of his wife, children, and domestics.

35 According to A. Irving Hallowell, a relatively unexplored territory is ethno-metaphysics, that is, the cognitive orientation in a cosmos of people in whatever culture even though the basic premises and principles implied do not happen to be consciously formulated and articulated by the peoples themselves. HALLOWELL (1960), p. 20: „Human beings in whatever culture are provided with cognitive orientation in a cosmos; there is „order" and „reason" rather than chaos. There are basic premises and principles implied, even if those do not happen to be consciously formulated and articulated by the peoples themselves. We are confronted with the philosophical implications of their thought, the nature of world of being as they conceive it. If we pursue the problem deeply enough we soon come face to face with a relatively unexplored territory: ethno-metaphysics!" See also FOSTER (1965), p. 293 ff., and the ideas of SCHUMACHER (1973), part 1, on the perception of nature in capitalism.

This older economics or *oikonomika* found its groundwork in Aristotle, in his *Politics* and *Nicomachean Ethics*. In the Aristotelian form it has determined the ancient European history.[36] It is a doctrine of the „whole house" as unity of living and working and a theory of one of the basic forms of societal organisations in all rural and rural-feudal cultures, of the rural *oikos* which, from its beginning in the neolithic period until the nineteenth century, formed the basis of the European social structure. The rural economy is defined by two structural traits: firstly, by the dominant autarky of the provision of subsistence and a small degree of production for the market, and secondly by the central direction of the economy by the master of the household. The household is, in the words of Walter Eucken, „a simple, centrally directed economy." The relationships between the master of the house and the familial and working members of the household are those of rule and status, not of contract. The ancient European *oikonomika* is a theory of the „whole house" as a unity of household and business.

Already in antiquity, the rural subsistence-economy has been superimposed by the town centered market economy with trade exchange, hired labour and a monetary economy. Beside the rural basis of the society exist „capitalist" overlappings. Aristotle has clearly seen this development towards capitalism and has refuted it. His critique can shortly be summarized like this: the new stage of the economy differs from the true *oikonomia* by that it does not aim at autarky and at satisfying a constant demand but at exchange value and at monetary profit. It makes humankind insatiable and legitimizes the boundlessness of the *pleonexia*, of the desire to have more, since the monetary wealth has, according to Aristotle who refers here to Solon of Athens, no limit that is recognizably set to humankind.

Aristotle names this form of the economy and of the art of making profit *chrematistics* after its goal, making money (*chremata*). The knowledge of chrematistics is not concerned with securing subsistence but with finding out how to exchange goods in order to gain more profit. Medium and means of the chrematistic way of conducting business is money, and its aim is again money. This inversion of goal and means and the endlessness of its needs and desires signify the unnatural character of the chrematistics. It hurts the basis of the Aristotelian ontotogy: measure and limited form. The means of human action, even the goals of business and of all other action can not be unlimited

36 Compare OTTO BRUNNER (1950) and KOSLOWSKI (1976).

and boundless for Aristotle. To strive for an unlimited goal leads into the void and is meaningless.[37]

In his critique of the chrematistic mode of economics and of business, Aristotle formulates the basic pattern of every critique of capitalism for which Marx praised him in *Das Kapital.* The attitude and science of chrematistics in which one has to see, with Otto Brunner, the proto-history of economics, violates the Aristotelian concept of a hierarchically formed, ordered and harmonic cosmos, because it allows for the boundless and infinite pursuit of profit. The chrematistic economic motive can not be fitted into the teleological ontology and destroys therefore the order of *polis* and *kosmos* and the analogous harmonious arrangement of society and nature.

In the development of social metaphysics and its correspondence of the perception of the order in society and in nature, the teleological ontology and social metaphysics in the line of descent of the *oikonomika* or household economics as a theory of the pre-capitalistic feudal economy can be distinguished from the mechanistic ontology and social metaphysics in the line of descent of the *chremanistics* or market economics as the theory of capitalism.

To the concept of *oikonomika* corresponds, then, the conception of the cosmos as a household, as God's *oikos.* Stoic philosophy saw the cosmos as a thoroughly teleologically ordered whole, as the great household of God, which He directs and protects as the good master.[38] The world is a great oikos consisting of mortals and immortals and directed by a master. The conviction runs from Homer to Thomas Aquinas: „But the things want to be governed not badly, but by *one*."[39]

The same structure appears in economics as in ontology: central government and subordination as principles of order. The world appears to prebourgeois or premodern man as the simple, centrally directed economy of God. Wolf Helmhard von Hohberg, one of the major authors of the *oikonomika* of the seventeenth century, writes in his *Georgica Curiosa* that God „is the charitable, heavenly house master, who never ceases cultivating and ruling over

---

37 ARISTOTLE, *Politics* I, 1258 a1 - 14, and *Nicomachean Ethics* IX, 1170 a 20. Cf. KOSLOWSKI (1979), p. 70, 80.

38 For example EPICTETUS, *Dissertationes* lib. III, cap. 24,92, ed. by JOH. SCHWEIGHÄUSER, Leipzig 1799, Reprint Hildesheim/New York (Olms) 1977, Vol. I, p. 509; further loci in SINGER (1958) and WAGNER (1969).

39 HOMER *Iliad* 2. 204. Homer's dictum is cited in ARISTOTLE, *Metaphysics Lambda* 1076a. 4-5; and Aristotle again in THOMAS AQUINAS, *Summa theologiae* 1 q.103a.3.

the great world economy".[40] The harmony of divine and human oeconomia is to him self-evident.

b) Teleology, Mechanism, and Preestablished Harmony

The image of the ordered world economy is inseparably linked with a teleological view of human nature in early European metaphysics.[41] As Leo Strauß writes, is the question of natural right decided by the way in which the problem of the motion of the celestical bodies is solved. His thesis must, however, be extended by the thesis that also the type of economic theory that someone holds true is closely linked to the viewpoint he takes with regard to the mechanical or teleological conception of the cosmos.[42]

For Aristoteles and the Aristotelian Middle Ages the circular motion, limited and returning in itself is the perfect motion. An endlessly forward going motion is inconceivable in a teleological ideology, in which every motion-even that of human action - is determined and limited by its characteristic goal. There can not be an unsatiable desire. Humankind's pursuit of happiness and material goods is on principle perfectible and satisfiable. The third structural characteristic of capitalism, the maximization of profit and utility under the assumption of unsatiable demand is senseless for Aristotelian economics since it presupposes an unlimited striving. In Aristotelian thinking, the theorem of micro-economics about the nonsatiety of demand can not be made the basis of a rational theory of human action.

With the advent of the civil society and its world view the concept of the teleology of nature and of the entelechy of every being consisting in its being perfectly active in the perfection proper to it is replaced by the mechanistic world view with its idea of universal mechanism, linear motion and infinite striving for self-preservation. Spaemann has demonstrated this „inversion of teleology" from the realization of the telos of a thing in its acting well to the purpose of mere self-preservation. He also analysed the rise and connectedness

---

40 Cited by OTTO BRUNNER (1950) p. 138. See for the medieval ontology KRINGS (1982).

41 Compare LEO STRAUSS (1953), p. 8, and H. GROSSMANN: „Die gesellschaftlichen Grundlagen der mechanistischen Philosophie und die Manufaktur", *Zeitschrift für Sozialforschung*, 4 (1935) pp. 161-231, reprint Munich (Kösel) 1970.

42 STRAUSS (1956), p. 8. Cf. also GROSSMANN (1935).

of bourgeois ethics and nonteleological ontology in B. Telesio, Th. Campanella and F. Bacon.[43]

The radical change to the modern anthropology and ontology becomes even more evident than in ethics in the economic theory of the beginning bourgeois era and of the first historical stages of capitalism. Hobbes and Mandeville do not any more represent the Aristotelian ontology and economics with their idea of a teleologically limited striving for satiety of needs. They put forward a mechanistic ideology and a „capitalist" economics. The motions in nature and society are thought to be such that if they are not hindered they preserve themselves linearly and endlessly. The endless and stimulating desires of men are not satiable. The self-preserving motions in nature and in the economy, the motion of the celestial stars like the activities of men must be mechanically coordinated.

Hobbes writes about the movement of human acting and the motivation-structure of men:

„Felicity of this life consisteth not in the repose of a mind satisfied. For there is no such *Finis ultimus* (utmost ayme), nor *Summum Bonum*, (greatest Good) as is spoken of in the Booke of the old Morall Philosophers. Nor can a man anymore live, whose Desires are at an end, than he, whose Senses and Imaginations are at a stand. Felicity is a continuall progresse of the desire, from one object to another; the attaining of the former being still but the way to the later".[44] The nonsatiety of human desires can hardly be formulated stronger.

The metaphysical-ontological presuppositions of the new society have been very evident to the proponents of a civic and bourgeois individualism. The connection between the teleology of nature and of human action, between the understanding of natural motions and that of human desire becomes evident in Mandeville's sarcastic remark about Aristotle's theory of asteroid motion:

„From what appears to us in human affairs, and the works of Nature, we have more reason to imagine that the desire as well as aptness of man to associate do not proceed from his love to others, than we have to believe that a

---

43 SPAEMANN (1963), p. 50 ff. This chapter is reprinted in EBELING (1976), where also the papers by H. BLUMENBERG and D. HENRICH on the relationship between selfpreservation, subjectivity and modern rationality are included.

44 THOMAS HOBBES, *Leviathan*, p. 160.

mutual affection of the planets to one another is not the true cause why they keep always moving together in the same solar system."[45]

According to Mandeville we can not assume like Aristotle that the stars move around each another in circles by their mutual attraction like loving people. Even less reason have we to believe that societal association and economic exchange come about from the love of humans for each other. The stars are kept on their course in the empty space by the infinity of the linear motion and by gravitation; society is kept together by the endlessness of human desires, the mechanism of mutual advantageous exchange, and the skilled management of the statesman.[46]

Hobbes and Mandeville create the metaphysical and anthropological reasoning and justification for the third structural characteristic of capitalism, the assumption of the nonsatiety of demand and the infinity of the profit motive. Their anthropology and social metaphysics, however, do not allow to develop a complete theory of the market as a co-ordination procedure. The complete rejection of teleological elements excluded the idea of an order spontaneously arising from the pursuit of indiviual goals by an invisible hand. There is a partial proof of the *laissez-faire* principle in Mandeville's principle of *private vices - public benefits* that sets free individual self-interest and releases economic needs and effective demand from moral norms and restrictions. Against Hayek's thesis that Mandeville is the founder of laissez-faire[47], however, the concluding sentence of „Private vices - public benefits" must be remembered: „I conclude with repeating the seeming Paradox, the Substance of which is advanced in the title page. That private vice *by the dexterous management of a skillful politican* may be turned into public benefits".[48] The model of pure mechanism in Hobbes and Mandeville is not the foundation of a market society, but of etatist mercantilism.

The mechanistic model of human nature freed humankind's commercial strivings because it interpreted desire as boundless and as the power which drove man. It was unable, however, to explain the *spontaneous concert* of actions in the market. The justification of the central structural characteristic of capitalism, of the coordination for the common good of the pursuit of indi-

---

45 B. DE MANDEVILLE: *The Fable of the Bees: or, Private Vices, Publick Benefits* (1714-1729), ed. F. B. Kaye, Oxford 1924, Vol. II, p. 178. Compare for Mandeville, KOSLOWSKI (1982), p. 174-184 and DUMONT (1977).

46 See also on the ontology of economics KOSLOWSKI (1991).

47 HAYEK (1969), p. 129.

48 B. DE MANDEVILLE, *op.cit.*, Vol. I, p. 369.

vidual goals by way of market and price mechanism, requires the assumption that forces towards equilibrium are active, that is, a teleological element in a mechanical system. The model of general mechanism can only assume a rigid composition of forces like in a clock-work, but can not explain the evolution of varying and spontaneous forces. The mediation between the mechanical and the teleological spontaneous model of ontology was accomplished in philosophy by Leibniz's concept of the preestablished harmony.[49]

In economics the model of a preestablished harmony of forces that lead in their concert to an equilibrium was introduced by Adam Smith's theory of the *invisible hand*. The mechanism model can only explain the mutual effects of utility maximizing individuals on each other, their exertion of force and counterforce, but not the emerging of a spontaneous order, which fulfills the optimality conditions. These are only fulfilled when a conception of preestablished equilibrium is added to the mechanism concept.

In so far Sombart's thesis[50] of the correspondence between the theory of Newton's celestial mechanics and the model of equilibrium in liberal economics is not quite correct for two reasons, a historical and a systematic one. Historically, the mechanism model has been adopted only in part by classical economics. Systematically, the combination of spontaneous forces cannot be explained by the mechanism model. Thus the Newtonian theory of the motion of the stars that are kept in an eternal equilibrium between repulsion and attraction and follow their course in wonderful harmony could be applied to human society by the physiocrats, but not by Adam Smith.

The physiocrats claimed that for human society, just as for the starry heaven, there is a preestablished order set up by God, which in its unimpeded workings brings forth the highest standard of happiness and prosperity. Just as the celestial bodies tend toward a harmony of the spheres, the free activities of the economic subjects lead, by means of a process of attraction and repulsion, to a harmony of interests on earth.

The theory of laissez-faire and of the market as a machine being constructed to the best by god, is so optimistically convinced of the wisdom and the power of the creator and of the perfect goodness of creation that, for ex-

---

49 Cf. G. W. LEIBNIZ: *Die Theodizee* (1710), tr. by A. Buchenau, Hamburg (Meiner) 1968, particularly p. 264. Cf. BONAR (1909), p. 221, and VINER (1972), p. 60: „Eighteenth-century British social philosophy was in fact soaked in teleology. I know of no British writer before Bentham who frankly renounced teleology".

50 Compare SOMBART (1938), p. 13, and, for the Leibnizian metaphysics of modern economics, KOSLOWSKI (1991).

ample, the elder Mirabeau can write: „All direction of the economy by the state is either according to the laws of nature, and therefore superfluous, or against them, and therefore without success."[51] Mirabeau demanded to popularize the laws of the natural economic order in all schools as the „Religion of the Bread of Everyday" (*Religion des täglichen Brotes*) and the theories of the physiocrats to be posted in all schools, in all city halls, and in all sacristies „as the object of an earthly cult and as a disinfectant against the rampant disease of inhumanity."[52]

c) Coordination without Domination: *katallaxia* and Deism

Despite the differences in their basic mechanistic or teleological models, the physiocrat and the Smithian models of the economy share a metaphysical characteristic: deism. Economics, thus, corresponds theologically to deism and ontologically to the model of preestablished mechanism.[53] The basic metaphysical principles of deistic theology and social philosophy are identical: absence of domination as opposed to central planning, noninterference as opposed to interventionism, and coordination as opposed to subordination. These are the standard characteristics of a market society. Just as little as God intervenes in the world, so should the state intervene little in the economy.

The paradigm of economics was transformed into the exchange model of economics, into catallactics. The economy as centrally directed household in which the use of resources serves a homogeneous hierarchy of aims is substituted by the spontaneous order of the market. The theory of *katallaxia* replaces *oikonomia*.[54], the model of the market substitutes the household model. The metaphysical changes lead from a dominating God to a deistic concep-

---

51 Cited by ROSCHER (1874) p. 481. Cf. also FREYER (1921), p. 14/15.

52 VICTOR MARQUIS DE MIRABEAU: *Letter to the Markgraf Carl Friedrich of Baden from 31st of March 1770*, ed. by Carl Knies, Heidelberg (Winter) 1892, Vol. I, pp. 22 and 24.

53 See BONAR (1909), p. 221.

54 *Katallaxia*, from *katallattein* (Greek) = to exchange, means the exchange model of the economy. According to SCHUMPETER (1965), p. 655, the term *catallactics* meaning theory of an exchange economy was introduced by Archbishop Richard Whately (1787-1863) to replace the term „political economy". However, Whately was not successful in doing so. The term has been taken up again by the Austrian school of economics by Menger and Gossen, based on their emphasis on exchange value, and by the younger Austrian school by VON MISES (1949, p. 233), and VON HAYEK (1969, p. 112).

tion of a God who holds himself aloof from the world. The political model changes from hierarchy and subordination to egalitarian democracy and coordination.

The conceptual difficulties that follow from the relationship between human freedom and the natural mechanistic order, on the one hand, and from the relationship between God and the world in deism, on the other hand, were much neglected during the Enlightenment and were first taken up again in the philosophy of German Idealism. In this process of secularisation economics itself turns into a piece of metaphysics and theodicy.

Leibniz used the notion of the preestablished harmony in the *Theodicy* in order to show that this is the best of all possible worlds. The theory of the harmony of interests in Mirabeau also contains a theodicy. In the process of secularization, God's economy of salvation is transformed into an economic history of salvation and a philosophy of history in Saint-Simon and Marx. Irenaeus and Tertullian introduced the concept of *oikonomia* or *oeconomia* into Christian theology as the scene for God's plan for saving mankind. They summarized God's incarnation for the salvation of man in the *economic trinity* and developed the doctrine of God and salvation into a functionally consistent dogma.[55]

With deism, a direct intervention of God for the sake of salvation is theologically excluded. Humankind cannot hope for salvation through the divine economy of salvation, but is dependent on the workings of the human economy within the world. Since the system of natural freedom and capitalism was unable to fulfil the utopian hopes for overcoming scarcity, the problem of the realization of the best of all possible worlds was transformed into the economic utopia of Karl Marx who had the illusion that the evils of the world could once and for all be overcome by an economic reorganization of the means of production. This economic utopia, too, has been disappointed by now.

55 Compare article „oikonomia," in: *A Patristic Greek Lexicon*, ed. G. W. LAMPE, Oxford (University Press) 1961.

Chapter III

Conceptions of the Market Economy: Models of Mechanism and Models of Interaction

In general, theories of capitalism are economic and not philosophical. The distribution of labour in the sciences has the result that the economists deal with the economic substructure and the philosophers with the superstructure of ethics and values whereas an integrated overall view is missing. This narrowing of perspectives leads to serious gaps in the theory of the capitalist society.

Within economic theory two approaches to the theory of capitalism can be distinguished: first the mechanistic model of general equilibrium in neoclassical economics and secondly the model of interaction in the Austrian school of economics. Hutchison distinguished between a „Smithian" and a „Ricardian" foundation of market economy.[56] The Smithian tradition formulates social theories in a broad perspective including the political, legal, and ethical framework of the market, whereas the Ricardian tradition analyses the social problem in a narrow economic perspective and emphasizes aspects of efficiency of an utopian optimism that require complete anticipation.

Typically the neo-classical model can be interpreted as a mechanism that leads to the equilibrium if allocation is efficient. The efficiency of the allocation of resources for the satisfaction of existing demand stands at the center of attention. Since freedom leads to the greatest total benefit all obstacles to freedom have to be eliminated, but if this allocative optimality is disturbed, other efficient modes of allocation have to be found. Thus the market, as Buchanan put it[57], is turning into a „mechanism, into a calculating machine that computes information and receives input or transforms output", a mechanism that, if it is disturbed (market-failure), is substituted immediately by other mechanisms of allocation like the state or voting.

On the other hand, in the „Austrian" tradition the market is a context for the interaction of acting individuals pursuing their individual ends. A context that mediates the individual persecution of objectives with those of all others and is thus facilitating freedom as the ability to act according to self-determined objectives. Next to the aspect of an optimal allocation the aspect of

56 HUTCHISON (1979), p. 433.

57 BUCHANAN (1964), p. 219. Cf. in the following KOSLOWSKI (1982), chap. 6.

freedom is gaining importance. As Buchanan writes[58]: „The market or market organization is not a means toward the accomplishment of any thing. It is, instead, the institutional embodiment of the voluntary exchange processes that are entered into by individuals in their several capacities ..., it is a setting, in which we observe men attempting to accomplish their own purpose, whatever these may be.“ According to Hayek[59] the market is that process of discovery in which we first experience what our objectives are and which resources are scarce.[60] Correspondingly representatives of this tradition are more readily prepared to stick to the market as an instrument of allocation even when forced allocation might be more efficient.

The problem of the ethics of capitalism arises within the economic theory in the openness of the system, in the possibility of a trade-off between efficiency and economic freedom.[61] The model of general market equilibrium has no unequivocal, automatic solution but rather remains dependent—because of market failure in the case of externalities, indivisibilities, and so on—upon the metaeconomic determination of the optimal relation between freedom and efficiency.

The model for the mechanism of choices leaves open in the last analysis a non-mechanical grade of freedom. How can one achieve a balance between freedom and efficiency in the economy? According to J. Marshak, „the sacrifice of liberty is an organizational cost“.[62] However, this does not advance the inquiry as to how this sacrifice of freedom is to be evaluated. This question remains one of a balancing between social values which transcend the purely economic model.

One cannot immediately switch to forced allocation at every disturbance of the optimal conditions nor can one, as K.J. Arrow correctly objects, hold non-interference to be the only value.[63] The science of choice, welfare economics, is of no help here either as the discussion of the possibility of a social welfare function has shown.[64] A mechanical solution of the overall optimum is not available.

F.H. Knight has critically examined the mechanistic analogy in economics. The general equilibrium model interprets economic behavior in ac-

58 BUCHANAN (1964), p. 219.
59 HAYEK (1968), p. 7 and 9.
60 Cf. also to LUHMANN (1968), p. 15 sqq.
61 Compare ROWLEY and PEACOCK (1975) and DUPUY (1978).
62 MARSHAK (1974), p. 199.
63 ARROW (1967), p. 12.
64 See BUCHANAN (1969), pp. 62-64.

cordance with the analogy of force in which the motive causing an act is understood as force.[65] Economics must then be concerned with actions arising from preferences that are not further questioned. The Newtonian concept of force in mechanics has been criticized in physics as metaphysical by E. Mach and H. Hertz, but it acquires justification insofar as forces in nature and observable and experimentally reproducible. That is not the case, however, for economic preferences, which are conceived as forces behind the choices of individuals.

Preferences have a primarily social character; that is, they are influenced by social status, training, and education, as well as by error. Preferences and choices are not identifiable like force and the effects of force in the natural sciences. Market competition cannot be considered according to the model of force and opposing force which lead to an equilibrium of forces. The mechanism model takes goals, motives, and the preferences of individuals for the given and only accepts considerations as to means. It is a model of adaption in which the individuals' acts of choice are based upon their unquestionably accepted preferences and the force conditions of the market. No value problem concerning the selection between goals in accordance with certain categories arises in this model.

Chapter IV

Morals in Capitalism, or: Are Morals Superfluous in a Working Market?

Through the acceptance of given, constant goals, the moral problem is reduced to an economic one, and ethics is replaced by economics.[66] Knight sees

---

65 KNIGHT (1935 c), p. 241 and (1935 d) passim. Cf. also KOSLOWSKI (1983 b).

66 KNIGHT (1935b), pp. 34-5, asserts, „The assumption that wants or ends are data reduces life to economics and raises again the question with which we started out: Is life all economics or does this view require supplementing by an ethical view of value?". Similarly, decision theory does not tell the decision maker what to prefer but what to choose under given preferences. According to STEGMÜLLER (1973), p. 325, „Normative decision theory is not an ethic". Thus, decision theory cannot give, in general, effective advice for

this „displacement of ethics by a sort of higher economics"[67] in classical economics and in utilitarians like Bentham and J.S. Mill, and it also appears in the case of Spencer[68]. Looking at contemporary theoretical discussions one could see sociobiology and bioeconomics as attempts to introduce a universal economics. For utilitarians, who are thoroughgoing hedonists, well-being is the goal of all action. Ethics is then reduced to the optimal allocation of resources for the goal of the greatest pleasure. This goal is empty and formal, and as long as the substance of pleasure is not determined, the maxim of maximization of pleasure means no more than each doing as that person wishes anyway. No help in making selections can be drawn from the concept of the greatest pleasure or utility. On this basis, the individual cannot choose if he does not already know what he wants. The „economy of pleasure" leaves the individual helpless if he did not know already at the beginning of his cost-benefit-analysis what kind of pleasures he is striving for.

Among the objections to hedonism that have been raised since Plato are that people do not want pleasure or utility but rather seek concrete goods and that they do not desire plain pleasure but distinguish and rank various pleasures.

Thus, Plato argues in *Gorgias* that a person interested only in satisfaction should best wish himself an itch, so as to be better able to scratch himself.[69] Max Scheler further criticized hedonism by saying that one cannot attain happiness immediately but rather obtain it „on the back of other activities".[70] One does not play the piano in order to be happy; rather, one can be happy when one knows how to play the piano. One's interest in capitalizing precisely does not obtain the goal when one does not do things for their own sake.

Knight[71] raises against hedonism the old philosophical and religious wisdom that suppressing desires instead of trying to satisfy all of them may lead to a greater degree of happiness and is definitely more economical. Finally, in accordance with Alchian[72], it can be refered to the fact that under conditions of a complex environment of an action and of uncertainty about the strategies

the selection between alternatives but mostly tells the decision maker only more precisely what he already knew.

67 KNIGHT (1935 b), p. 19.

68 SPENCER (1982).

69 PLATO, *Gorgias*, 464c.

70 SCHELER (1966) p. 351.

71 KNIGHT (1935 b), p. 31.

72 ALCHIAN (1977), p.17.

of the others, utility maximization cannot be considered as an operational advice of action since the individual cannot find out which of all possible courses of action is exactly utility maximizing.

The two approaches just presented, that is, the elimination of ethics from capitalism by means of the mechanism model and the attempt of utilitarianism to make economics into a kind of higher ethics, share the intention of reducing incommensurable preferences and needs to a common denominator. Within the mechanism model this is effected by having recourse to constant motives as the relevant acting forces which reduce the problem of choice to a problem of adaptation, within utilitarianism by tracing back all motives of action to the formal motive of utility maximization.

Both approaches are attempts to avoid the value problem inherent in the selection among goals by using tools of economic theory. Thus, „economics might almost be defined as the art of reducing incommensurables to common terms. It is the art of heroic simplification".[73] Both positions amount to the attitude correctly described by K.E. Boulding as „knowing the value of nothing and the price of everything,"[74] which means nothing other than that the economy can show the individuals the relative prices and the optimal allocation of their resources for certain goals, but cannot relieve them of the choice between goals and values.

1. Formation and Coordination of Preferences: The Coherence of Ethics and Economics

Preferences are not rigidly given and invariable, and the social problem is not merely one of economizing the use of means. Preferences are ethically and socially transmitted; they are formed in individual ethical reflection as well as social interaction. Symbolic interactionism, as was presented by G.H. Mead and W.I. Thomas shows how closely one's view of the world and one's perspective is determined by groups and communities.[75] One sees goods not in themselves but rather in a close weave of perspectives of different reference groups to which one belongs and in the symbolic definitions one gives to the qualities of the goods.

---

73 SHACKLE (1972), p. 10.

74 BOULDING (1967), p. 67.

75 According to KNIGHT (1935c), p. 129, symbolic interactionism „offers much greater possibilities for throwing light on behavior than does the science of behaviorism" .

According to the Thomas theorem, symbolic definitions of situations that people adopt are real in their consequences. From this point of view, culturally defined needs are as real in their consequences as physiological needs. One can consider needs as constant only in abstraction and in the very short run. De facto, preferences continually change by way of the transformation of institutions and society.

As deliberation upon the correct allocation of resources for given purposes, economics can provide information on the possible extent and opportunity costs of the fulfillment of goals. However, Knight points out that it can „never get beyond the question of whether one end conflicts with another end and, if so, which is to be sacrificed".[76] In the case of a conflict between competing goals one must abandon the level of scientific economics and use preference rules. Ethically speaking, the question as to which goals an individual in a society sets for himself is more important than the question of how this goal is to be fulfilled.

It is obvious that the ethical and sociological theory of the formation of preferences is logically prior to economics as the theory of the allocation of resources for these preferences. One must ask about the reasonableness of the goals *and* about the optimal allocation of resources for these goals. Neither of these questions can be reduced to the other. Society is not conceivable as a pure *katallaxia*, as the problems of a Paretian social overall optimum show. With a given factor endowment and given preferences, a condition is conceivable in which the complete variability of all quantities and of the anthropological presuppositions for the indifference curves leads to a situation in which no one can improve his or her position without another being hurt. Can this be interpreted as a real optimum? Let us set aside the problem of the initial distribution of endowments for the moment. It is still apparent that all adaptations in the system are more or less of a strategic character. Preferences were not examined or transformed for rationality or goodness but under the assumption that they are fixed and given merely adapted to variations in the environment. The Pareto-optimum cannot, therefore, define social or ethical optimality beyond the economic viewpoint of allocative optimality.

The ethical postulate must still be raised that individuals should vary their effective demand not only according to their own given preferences and accommodate to those of others in the course of exchange in such a way that a Pareto-optimal position is reached, but that they should *transform* their preferences in an ethical way respecting the preferences of others. They should

---

76 KNIGHT (1935b), p. 37.

not only move on the indifference curve but change their system of indifference curves at times.

K.E. Boulding[77] claimed that every movement on the indifference curve to a point on the contract curve presupposes a certain indulgence and the absence of jealousy or envy. If one follows this thesis, the Pareto-optimum implies a moral minimum. Yet such a moral minimum in a Pareto system cannot be an optimum in a moral sense.

It is an underestimation of ethical reason and a scientific fallacy to assume that people only *adapt* their preferences *strategically* and are unable to *transform* them with respect to the totality of conditions. The model of the market and the Pareto-optimum acquire an ideological character when they are presented as the last word concerning the theory of action.

This is even more so the case with respect to attempts, by way of a recourse to sociobiological and bioeconomic categories, at saving *economics as the universal science* in view of the objections of anthropology and sociology which point to the cultural and institutional formation of preferences and choices. For example, J. Hirshleifer after having conceded that man being „full of love and hate and sheer cussedness ill fits to the model of economic man," introduces the economic gene in order to save the universality of economics.[78] The gene directs the choices of man so that he produces or promotes identical duplicates of his genetic equipment.

Clearly, the interest of the social biologist is to eliminate acts of choice in a strict sense. No longer the man but rather the economic gene is the doer behind the deed who attempts to strategically secure identical copies of itself. Here the choice between competing values is eliminated; there can no longer be any choice between conflicting goals. The unitary goal of nature and society is simply given and all actions are merely strategically resisted to it. Apart from the animistic element in this theory and the conflicting empirical fact that with sinking birthrates especially wealthy societies do not contribute to the maximization of identical gene copies, although they could do so best, the theory of sociobiology presents the most reductionist interpretation of the value problem. However, sociobiology, contrary to its critics' assertions, is not a theory of capitalism since capitalism does not restrict itself to purposes

---

77 BOULDING (1973), p. 113. Following Takeo Doi, the thinking in terms of indifference curves presupposes that the economic agents practice some appreciation of the other's interest, practice what the Japanese call *amae* , the desire to be loved by others. Compare TAKEO DOI (1981).

78 HIRSHLEIFER (1978), p. 240.

of genetic reproduction but rather admits all individual purposes biological, economic, and cultural.[79]

The value problem in capitalism arises because it is not centrally predetermined, and individual evaluation of goals, that is, freedom, is held to be a value. Freedom is both a fact - as non-intervention - and a value category.[80] It can be seen as an instrument for reaching other goods as well as being a value in itself.

The understanding of freedom as pure noninterference with market forces eventually leads to a pure mechanism in which everything that occurs without political intervention appears to be good. This understanding of freedom can be seen in Spencerism and social Darwinism. In western societies, freedom, understood as the ability to act according to self-chosen goals, is seen as an intrinsic value which has pushed other values, such as stability, calculability, and personal continuity into the background. The concept of freedom is bound up with the dialectic of freedom and self-responsibility, without which a free capitalistic order is unthinkable. This burden of freedom brings about the fact that a free economy cannot be, as it were, derived from efficiency criteria, but rather presupposes a moral will to freedom. Freedom and property must be wanted, as an act of the will.[81]

The value of freedom as an intrinsic good comprises a certain fixed relationship to other values. They ought to be able to be valid only if they are freely affirmed by the individual, and only in the case, where the possible coexistence of the values that the individuals set for themselves is guaranteed.

Capitalism as the system of contractural labor relations and free entrepreneurship excludes the finalization of the economic process according to centrally set goals, that is, the household model of the economy. It does not eliminate the problem of value; it rather puts the burden on the individual. Economic individualism therefore, is, necessarily tied to ethical individualism.

The problem of economic individualism is the following: What must I do in order optimally to reach my goal under the given constraints, and under the condition that others pursue their goals? The ethical problem is, What should I want? What are reasonable preferences? The answer of ethical individualism is outlined in the *Critique of Practical Reason* by Kant, who must be considered the founder of the philosophical ethics that correspond to a market econ-

---

79 For a discussion of sociobiology compare below Part Two.

80 See KNIGHT (1947a), p. 4 and (1947b), p. 372.

81 See W. RÖPKE (1949) p. 280.

omy: „So act that the maxim of your will could at any time serve as a principle of universal legislation."[82] This ethic corresponds to the structural characteristics of capitalism: individualization, autonomization, and universalization.[83] It is, as in the case of a catallactic economy, not finalized but mediates between formal and individual goals. It attempts to provide a criterion, according to which it can be determined whether or not one's individual goals are compatible with those of all others. If the individual goals pass this test, on the next level the question arises: How can these goals coexist with each other in the reality of the economic sphere?

The same objection to Kantian ethics has been made as was raised against the Pareto criterion. It is not capable of providing an answer to the basic problem of practical philosophy: What should we do? The categorical imperative is more a criterion of negative rejection than a procedure for the selection of goals.

One cannot live without values as criteria for deliberation between alternative courses of action. In this case, the bounds of rule-directed, methodical thought appear. The good can neither be determined solely as universalization nor expressed as a single category of value. In the selection of alternative forms of action one remains tied to the totality of the situation and of value categories, and one must treat the situation for action as a whole. It is one of the basic insights of philosophy that the good cannot be expressed in a single principle or value, but rather it must be determined as the totality of aspects of a situation and the nature of the matter. The idea of the good, according to Plato, cannot be presented as one single principle but rather as *truth*, that is, as the emergence of that which a person or a thing can be if it realizes the possibilities inherent to its nature *(aletheia)*, as *beauty*, and as *measure.*[84] Such a notion of the good certainly invokes methodological objections from a scientific point of view, to which Knight replies:

„This is of course intellectually unsatisfactory. The scientific mind can rest only in one or two extreme positions, that there are absolute values, or that every individual desire is an absolute and one as 'good' as another. But

82 KANT, *Kritik der praktischen Vernunft*, § 7.

83 Kant had a thorough knowledge of Adam Smith's work and a high esteem for his theory (KOSLOWSKI 1981, chap. 5). The structural similarities between Kant's and Smith's theories were stressed and criticized by the German conservative FRIEDRICH JULIUS STAHL (1802-1861), *Die Philosophie des Rechts*, 5 Vols., 1878, reprint Darmstadt 1963. In Vol. 2, p. 100 he sees in Smithonian economics the „analogon of Kantian social philosophy."

84 PLATO, *Philebos*, 65a.

neither of these is true; we must learn to think in terms of 'value-standards' which have validity of a more subtle kind".[85]

The difficulty of making such an ethic of the balancing of the totality of aspects operational puts quite a burden on the moral imagination of the individual, but the ability to deceive oneself as to possible relevant aspects of an issue to be decided is one of the characteristics of the immoral. The obligation to consider the totality of perspectives can compensate for the lack of operationality by the absence of a dogmatism of rules and a better consideration of possible side effects.[86]

2. The Need for Business Ethics

Especially within capitalism leaving a large space to individual freedom, there is the need for an ethic of the economy and of business that integrates the totality of value perspectives into the positive and normative analysis of economic choices. The need for a business ethic cannot be denied by contending that there is an automatism of competition.

The belief in market automatism is expressed by the frequently heard proposition that the market is not reconciliable with non-economic, ethical rules of behaviour. The dogmatic belief in the forces of market competition can be best illustrated by W.J. Baumol's dictum that the market cannot sustain a non-economic ethics and that the automatism of competition renders an ethics of business superfluous.[87] Under the conditions of perfect competition, voluntary moral acts of a single entrepreneur - such as ecological measurements, training for handicapped, and so forth - are not desirable for Baumol since the moral entrepreneur will be thrown out of the market within a short time. According to Baumol, „The merciless market is the consumer's best friend".[88] Voluntary supererogation only hurts the single businessman. Baumol asserts that social measures should be enforced by government on *all* firms:

Firms should not be „all-purpose institutions, but make money for their stockholders. [...] The notion that firms should by themselves pursue the ob-

85 KNIGHT (1935b), p. 40.

86 Compare for the value problem CHURCHMAN (1961) and on limits of rationality SIMON (1978).

87 BAUMOL (1975), p. 46.

88 *Ibid.*

jectives of society is, in fact, a rather frightening proposition... Corporate management holds in its hands enormous financial resources. [...] I do not want management to use the capital I have entrusted to it to impose its notions of international morality on the world."[89]

Baumol's dislike of moralism in corporate management derives from an unjustified optimism about the functioning of the mechanism of competition. Only if all firms produced on the break-even point and there were no producer rents, would every producer who fulfills supererogative actions be driven out of the market place. This case is very unlikely. Producers who produce with costs below the break-even point can afford moral voluntarism very well. In oligopolistic markets, managers can trade off slack for profitability.[90] They can choose between an easy life for the management and profit maximization. Therefore, they can choose between slack and moral actions as well.[91] Baumol's position shows the fallacy of the mechanistic model. In reality, the alternatives for actions allowed by the economy are much more complex than the classical cost minimization or profit maximization model suggests. Economic practice always takes place in a social totality in which the consideration of additional aspects of economic action which transcend the model of economic man is not only moral but can be profitable as well. Moral actions can have spill-overs in profits.

There is a certain irrational passion for dispassionate rationality in the economic theory of capitalism which bans any kind of moral motivation or thinking in terms of values from social science. It endangers the conditions of conservation of economic rationality.[92] As an example take A.A. Alchian and H. Demsetz's theory of the entrepreneur as the monitor of firm members' shirking: „Every team member would prefer a team in which no one, not even himself, shirked. Then the true marginal costs and values could be equated to achieve more preferred positions .... Obviously the team is better

---

89 BAUMOL (1975), pp. 46-47.

90 Cf. WILLIAMSON (1977) p. 188.

91 Baumol's position falls under KNIGHT'S (1935 b), p. 73, critique: „The striking fact in modern life is the virtually complete separation between spiritual ethics which constitutes its accepted theory of conduct and the unethical, uncriticized notion of efficiency which forms its substitute for a practical ideal, its effective values being accepted unconsciously from tradition or the manipulations of commercial sales managers".

92 Compare WEISSKOPF (1971).

with team spirit and loyalty, because of reduced shirking—not because of some other feature inherent in loyalty or spirit as such".[93]

Here the economist's belief in what Dennis Robertson calls the „need for economizing on love" goes too far. Critique of ideology adopts an ideological character itself when it denies the intrinsic value of moral action or constructs an opposition between morality and advantage, which ethical theory always has doubted. The ethics and particularly the natural right tradition has always—with the exception of Kant—claimed that there is an accordance of morality and *enlightened* self-interest. This accordance has recently even been confirmed by economic theory.

Sauermann and others have proven that trust reduces bargaining costs.[94] J.M. Buchanan shows that ethics is a substitute for the direct control in large groups where this control would be costly. Ethics is designed to solve the large number dilemma.[95]

As Sen has demonstrated, the general belief in ethical norms can solve the isolation paradox that each person would do (the) good if he knew that the others would do so as well, but will not do it if he might be the only moral individual in the group.[96] It changes the prisoner's dilemma into an assurance game.

Arrow interprets moral codes as reactions of society to the rise and for the compensation of market failure since moral codes can lower transaction costs and thus leave everyone better off.[97]

E.C. Banfield shows that the absence of trust and social integration and the predominance of non-enlightened self-interest is an obstacle to economic growth.[98]

A generally accepted business ethics lowers the transaction costs of market interactions. Transaction costs cause the shift of transactions from the market into firms, which form internal markets for labor and capital (m-form structure, holding corporation). These internal markets are more integrated than the general market by corporation-specific behavioral codes (corporate ethics, corporate philosophy). They increase the conformity between firm members' actual and expected behaviour and reduce transaction costs.[99]

---

93 ALCHIAN/ DEMSETZ (1977), p. 101.
94 SAUERMANN (1978). Cf. ALBACH (1980), p. 3.
95 BUCHANAN (1965), p. 8; (1978), pp. 364-8.
96 SEN (1967) p. 112; (1973), p. 119.
97 ARROW (1971), p. 22.
98 BANFIELD (1958) p. 89.
99 WILLIAMSON (1977) and (1981).

Even the mechanistic model of general equilibrium theory of neo-classical economics does not render ethics as the trans-economic evaluation of alternative actions superfluous. To decrease transaction costs, moral rules ought to be made generally binding and be internalized by the market participants. Micro-economic theory shows the need for a business ethics and for reembedding business, the market and the economic motivation into ethical and social norms. Thus, the structural characteristics of capitalism should be reembedded in the ethics and culture of a society.

Chapter V

The Morality of Capitalism and the Limits of Its Justification

According to R.A. Posner, „in a world of scarce resources, waste should be regarded as immoral".[100] Posner's statement could be read as a tautology with which everybody agrees: Waste is a pejorative notion, and no one would call waste moral. In the context of positivist economics, however, this sentence stands for a tendency to regard the allocation problem as the only ethical-economic problem since ethical judgments concerning goals are considered unscientific. In this perspective, capitalism would be the most moral system since it undoubtedly solves the allocation problem with the least waste as compared to other systems. Nevertheless, one cannot stop asking questions at the point of the allocation problem as has been demonstrated in the discussion of individual ethics.

1. Selectivity of the Market and Distributive Justice

For a justification of capitalism, the distribution resulting from an optimal allocation of resources must be investigated, as well as the question of whether capitalism selects or filters out certain goals in the market process. The apologists of capitalism have continually attempted to evade both questions by representing allocation and distribution as simultaneous processes

100 POSNER (1977), p. 23.

(marginal productivity theory) and disputing the selectivity of the market in pointing out that everyone, according to his willingness to pay, could realize all of his goals in the market. Both arguments are correct but not the whole truth. The arguments are connected. It is correct that allocation and distribution must go together, for otherwise there would be no incentives for an optimal allocation. Moreover, the positive contribution of a productive factor to the total product is one criterion of just distribution, and the consumer's willingness to pay for a given good is *one* standard for the intensity of preference. Goods should go to those who desire them most intensively, and because this can hardly be determined other than through willingness to pay, therefore, to those who are prepared to pay the most.

A competitive market does lead to the employment of every productive factor when it can bring forth the greatest product, measured in prices, and leads to a distribution which reflects productivity and relative scarcity. This argument from efficiency is not a sufficient ground for the morality of the distribution which results from the remuneration according to the marginal productivity of a factor of production. Even if the problem of economic computation could be solved, the problem of moral computation would remain.

All property rights on resources, whether labor (human capital) or capital in general arise from three sources: effort, inheritance, and luck.[101] Of these, only the first source can doubtlessly be called just, the second is merely legal, and the third incommensurable with justice. Thus, the distribution which arises from these three factors cannot be considered moral in an emphatic sense, but only be considered not immoral. Scarcity rents stand for the accidental characteristics of the distribution in the market process. Certain factors are scarce, given to their owners only by an accident of nature, yet are in demand. Other factors are just as scarce but not demanded. Is this sufficient reason to justify the enormous difference in distribution between both of the owners?

Hayek's[102] and R. Nozick's[103] approach that the distribution must be accepted as the result of a game that proceeds according to impartial rules and cannot be manipulated leaves the moralists unsatisfied and cannot even please the players. After a certain point in time each game requires a new dealing of the cards, a re-creation of the same initial conditions. For the game of life, which we can only play once, this must be even more true.

101 KNIGHT (1935a), p. 56.
102 HAYEK (1977).
103 NOZICK (1974).

A continual, periodic equalization of the initial positions is for reasons of efficiency not possible and criteria for such a distribution are lacking. Natural differences cannot be redistributed. But one cannot, as Nozick has suggested[104], disqualify every conception of end state in a theory of justice with reference to the rules of the game. That would be the capitalistic reversal of the dictum, let justice be done, though the world perish (*fiat justitia ut pereat mundus*), into the principle that the rules of the game must be obeyed, even when the chances in the game are very uneven, and even when the end results can be predicted and are trivial.

A purely deontological entitlement theory of justice is as abstract as a consequentionalist theory of end-state justice that continually shapes society in accord with its image of end-state justice. Hegel's remark in his *Philosophy of Right*[105] on ethical principles applies to both: „The principle of scorning the consequences of action, and the other, of judging actions by their consequences and making them the standard of what is right and good are both abstract understanding."

The unquestioned acceptance of the primary distribution which results from the market in capitalism without giving consideration to the final social effects of the effects of the economic process is no more moral than an arbitrary redistribution that is continually reshaping society and economy according to a prefabricated image of social justice.

Since distribution and selectivity of the market are correlated, both problems reinforce their respective advantages and disadvantages. An economy solely grounded on marginal productivity and effective demand results in an unequal distribution and means that the wealthy succeed better in enforcing their goals in the market. Nevertheless, this inequality can be compensated for by the fact that the consumer needs of the lower income classes become more homogeneous and, therefore, can be satisfied at lower relative prices if in the production of mass goods economies of scale appear.

Problematic, however, is the position of the consumer in the market who has a low income and particular or even extravagant preferences. He will find himself in the position described by Werner Sombart: „What use is it to me if I am a friend of hackney cabs, or of stove heating, or of silent film? I do not have any other choice but to consume automobiles, or central heating, or sound film, or - passive resistance."[106]

104 NOZICK (1975), p. 144.
105 HEGEL, *Philosophy of Right* (1821), § 118.
106 SOMBART (1937), p.16.

The same is true for the position of those producers who have a valuable good to offer for which there is, however, no effective mass or elite demand. The extremely subjectivist theory of value, which would hold that under these conditions the good in question is not really a valuable good, ignores the reality of the market, which is always determined by the whims and ignorance of the public.[107]

Especially in more egalitarian capitalist countries where certain cultural goods have not been reached by the public demand yet, and where those classes which so far have made possible the production of certain superior goods by their higher willingness to pay are due to this egalisation not able to do so anymore, a specific form of market failure may result. This is true, for instance, for certain technical and craft abilities which have been driven out by the industrial mass production. Already at the end of the eighteenth century, defenders of the guild system like Schlosser[108] have pointed to this process of driving out of crafts by industrial mass production.

Industrial mass production and the buying power of the majority taste cause an egalitarian tendency in capitalism that is often overlooked - despite all attention to developments in income distribution. It causes a tendency for the assimilation of equal life-styles behind the unequal income distribution. This tendency stands in contrast to the fact that one of the main and correct arguments for capitalism is that it makes individual life-styles and ways of life possible to an extent no other economic system did or does.[109]

Consumer sovereignty and the direction of market production by subjective preferences and effective demand causes the Janus-headed character of capitalism between egalitarianism and luxury and between freedom from domina-

---

107 „Giving the public what it wants usually means corrupting popular taste", KNIGHT (1935a) p. 57.

108 JOHANN GEORG SCHLOSSER: *Über das neue französische System der Policeyfreyheit, insbesondere in der Aufhebung der Zünfte*, Kleinere Schriften 1. Theil (Translation: About the new French system of police freedom, especially in the abolition of the guilds, Smaller writings, part 1), Basel 1787, reprint New York/London (Johnso) 1972, pp. 138-144. Schlosser closes with the sigh:

> „A, Seigneur, par pitié
> Pas tant de concurrence et plus de sûreté!"

109 MACRAE (1981) stresses the possibility of individual life-styles in capitalism.

tion and the oligopolistic power of the big firms. The fixation of prices according to demand and supply only removes all feudal privileges and causes a sociological equalization. On the other hand, it sets preferences and demand free of all social restraints or aesthetic criteria and allows parvenu luxury just as well as personal austerity. The market removes social barriers, on the one hand, and creates economic inequality, on the other hand, by the enormous accumulation of wealth as rewards for successful innovations or as quasi rents.

The tolerance of capitalistic societies toward the accumulation of wealth and the lacking embeddedness of wealth in ethical or social norms is certainly one of the critcrical points of capitalism. It makes the need felt for a new awareness of legitimacy and social obligation among the economic elites.

Economical inequality raises the question of power. To what extent are individual preferences effective within societal and economic decision-making processes, how far are these processes distorted by economic power in favor of certain groups? In order to answer this question, it would be incorrect to compare the really existing, imperfect markets with ideal, perfect procedures of democratic participation. Only realizable conceptions of decision-making processes may be compared to each other.

With respect to the problem of power, Buchanan[110] has referred to the structural relationship between the market and voting, that is, between the dollar vote and the equal right to vote at the ballot. Markets and democracy both require that decisions be made on the basis of individual preferences. As the Condorcet-Arrow paradox has shown, the market or the dollar vote is much more successful than voting in doing so. The market can register the intensity of preferences which are expressed by the individual's willingness to pay, and it forces the individual who must cast the dollar votes to a greater expression of his or her preferences and to greater responsibility in his or her decisions. The market increases by its continuity and non-restrictedness of decision alternatives the individual's chances to participate in decision making.[111]

---

110 BUCHANAN (1954a and 1954b), BUCHANAN/ TULLOCK (1962).

111 WALTER HELLER cites in: „Can Capitalism Survive?“, *Time-Magazine*, July 14, 1975: „The private market makes trillions of decisions without any central regulation. It is a fantastic cybernetic device that processes huge amounts of information in the form of the consumer voting with his dollars, the retailer telegraphing back to the wholesaler, the wholesaler to the producer“.

2. Limits of Subjectivism in the Theory of Values

Nonetheless, there remains a feeling of moral dissatisfaction with the market mechanism that cannot be explained merely by the inequality of the initial distribution and, therefore, by the greater choice possibilities of inherited wealth. For the market to be moral and the point it chooses on the production possibility frontier to be reasonable, would require that the effective demand, that is, the preferences of consumers were moral and their knowledge perfect. No one could assert that the choice decisions in the market are on the whole ideal or reasonable. Too much nonsense, bad taste, and superfluous luxury wins out over necessary, meaningful, and beautiful goods.

In addition, not only are given preferences coordinated and streams of factors directed through the demanded production of goods, but new needs are created through the market as well. As Knight points out, „The economic system forms, transforms, even creates wants. An examination of the ethics of the economic system must consider the question of the kind of wants which it tends to generate or nourish as well as its treatment of wants as they exist at any given time".[112] The moral guilt for many nonsensical needs is not to be borne alone by firms, which want to introduce new goods, but rather by the drive to imitation and the prestige needs of consumers.

It is thus rather remarkable that the proponents of economic democracy criticize the capitalistic system. For if consumers are incapable of asserting their sovereignty as consumers who are opposed to commercial advertising, they cannot begin with the presupposition that the choice makers in a democracy would be able to maintain their choice sovereignty with respect to political advertising in a plebiscitary democracy. Both kinds of „people's sovereignty", the sort of consumer sovereignty which is detached from all religious, aesthetic, and moral norms in a market system which is *solely* oriented toward subjective needs and the sort of political people's sovereignty detached as well from all norms in a plebiscitary democracy without constitutional norms, are equally a cause for fear.

The market system has always brought forth the criticism from religious quarters which expressed the opinion that the market system gave too much room to human irrationality and overly trusted the ability of human rea-

112 KNIGHT (1935a) p. 46.

son.[113] In contrast to the interpretation of Edward R. Norman and J.V. Schall[114], according to which capitalism is conceived upon a pessimistic anthropology based on original sin, the conception of a subjectivist and domination-free market represents the result of an optimistic, enlightened anthropology, as was shown previously in the case of Mirabeau.

One cannot dismiss the religious critique as paternalism. As Max Scheler asserted, „We all believe either in God or idols".[115] When the religious formation of preferences is abandoned, other forces for molding preferences, other Gods or idols appear.

A pure catallaxy is an ideal of coordination but not a meaningful social program because one does not come into the world as a freely decision-making being and utility-maximizing economic (hu)man, but is rather subjected to the influence of the environment and of the social reference groups to which one belongs. One needs the institutions and the action-directing norms just as much as, occasionally, paternalistic direction. The theory of market failure and of free-rider behavior with respect to public goods shows this. A society built only on revealed preferences would not be a nice place in which to live.

When under certain conditions the methodological individualism of the market leads to fallacies of composition, that is, the fallacy that what is good for the individual is also good for the whole, then this should be true of the voting process as well. Under these conditions, nothing is to be gained by the transfer from the market to a plebiscitary democracy.[116] It is much more to be feared that the needs which are not fulfilled in the market (public goods, cultural goods, the environment) would not be properly considered in an ideal democratic process as well. A good portion of the critique of capitalism is equally a critique of democracy and a critique of the inability of individuals to make reasonable use of their consumer and voter sovereignty.

The strengths of capitalism, that it can admit of many goals and of many values insofar as they can be borne by the market and that it abandons the attempt to finalize social and economic processes, are weaknesses in the eyes of those who hold that the market does not properly deal with certain values.

113 O. VON NELL-BREUNING (1974, p. 120), one of the exponents of Catholic social thought in Germany, admits, however, a certain „blindness of theology" towards the efficiency and possibilities of markets.

114 NORMAN (1979) p. 10, and SCHALL (1979) p. 37.

115 SCHELER (1966).

116 Compare KOSLOWSKI (1982), final chapter, and (1983 a) for the problem of market and democracy failure.

Criteria are, however, not available which would tell in which succession or in which intensity values and goals should be realized by the economy. Because freedom in the first place means the ability to set goals for oneself as Kant has demonstrated, one must concede to the economic actors the freedom to set goals for themselves even if one knows better in which order these goals should be realized.

Nevertheless, objections must be raised against the *value agnosticism* of capitalism which contends that one does not have available any criteria at all. Freedom cannot be the only value that a society can further. One cannot hold an allocation mechanism to be moral which, as Malthus writes, „denies a man a right to subsistence when his labour will not fairly purchase it.“ The justification for the non-finalization of the economy in wealthy countries can only go as far as the needs of subsistence have been secured.

3. The Significance of the Non-Finalisation of the Economy

That capitalism, despite its successes, has such a bad reputation is to be regarded as a result of the not understood problem of the non-finalization of the economy. Almost every group considers its own goals to be insufficiently provided for by the market system because the goals of the market system cannot be fixed by the group in question. Farmers see their market results as being insufficient, as do artists and philosophers. That the intellectuals are especially active in the critique of capitalism is caused, as Norman has clearly shown for the English tradition of capitalism critique[117], by the fact that they view their goals as not being sufficiently encouraged because of the lack of mass demand for what they supply. This is certainly the case for representatives of the social sciences. Capitalism with its trust in spontaneous, unplanned order offers them little opportunity for the implementation of their knowledge. In this respect capitalism contrasts with planned economies, which by definition must make the social planner the director of the economic process. The effects of this phenomenon on the income and wealth distribution are even favouring an increase of the critique of capitalism among the intellectuals and the intelligentsia.

It can not be denied, on the other hand, that capitalism also favors a certain group as far as the distributional results are concerned, namely, the group

117 NORMAN (1979).

that fulfills best the conditions and expectations on which its economic system is based: the group of the successful entrepreneurs. Nevertheless, it severely punishes the same group if it does not fulfill the system's expectations: the unsuccessful entrepreneur.

Every conceivable system always favors the type that corresponds best to its definition. Planned economies favor the planner, theocratic societies the priest, and belligerent societies the military. Capitalism has here the advantage that the fulfillment of system and role expectations and the social and economic remuneration of the economic elite are fairly efficiently connected with the interests and needs of the population, that is, with consumer demand. The consumers' possibility for „exit"[118] in a competitive market and the control of profits by competition between producers assure to a certain extent that economic success and remuneration are linked to socially desirable and useful performance.

Against the critique of interest groups upon the allocation and distribution effects of a nonfinalized capitalistic economy, one must recall one of the oldest views of justice in the European tradition, the idea of equilibration[119] and measure. A society which bases the pursuit of its goals on *one* normative principle only, or supports the purpose of only *one* group is not able to realize the good.[120]

The idea of an equilibrated totality of goals and ends is an idea that must be aimed at the critics of capitalism as well as at those who absolutize the allocation mechanism of the market. The theory of market failure as well as that of government failure indicate that a balance must be found between society and state, market and voting.[121] Both the market mechanism and the state show their failures. The morality of capitalism cannot consist, as Knight[122] has shown, in introducing abstract economics as absolute ethics, that is, in reducing all questions of social and ethical values to the question of the optimal allocation of resources for satisfying given individual prefer-

---

118 HIRSCHMAN (1970).

119 For the importance of the concept of equilibration (*Ausgleich* in German) see KOSLOWSKI (1996).

120 See for the tradition of the mixed constitution and separation of powers FRITZ (1954), p. 490.

121 The dualism of state and society is a structural constant of European societies since the Greek polis. Compare KOSLOWSKI (1982). The limits to the discourses of market and voting are discussed in KOSLOWSKI (1987).

122 KNIGHT (1966).

ences. The totality of the social order must not be understood only as a market.

The moral justification of capitalism consists rather in mediating many goals and their pursuit by individuals in such a way as to preserve moral and economic freedom without a war of all against all. That which the individual and society takes to be preferred can only be reached by the market through a compromise between what the individual takes to be important and that what all others take to be important. A compromise is all that can be reached when individual pursuit of goals is allowed.

Competition in the market is and must be a competition ruled and limited by law. It is - using a risky paradox - a kind of institutionalized peaceful civil war. „In case it (competition) goes out, we are threatened with a kind of pensioner's existence at the expense of the state, whereas the external competition between the states persists. Terror fills this gap", Ernst Jünger asserts.[123] The terror of social determination of goals by *one* group is the alternative to competition in advanced societies because, according to historical experience, only through terror can a society which has experienced freedom be sworn to a particular goal. This interdependence has been proven once more by the experience of the East European planned economies.

Chapter VI

Some Social-Philosophical Conclusions

1. Capitalism as Utopia

According to our knowledge of history, a purely capitalistic society, built exclusively upon private property, maximization of utility and profit, and coordination by way of the market and price system, has not yet been a reality. As a societal model, capitalism bears utopian, contrafactual features; it is itself a social utopia. Its utopian character always becomes evident when its defenders seek to immunize their opponents' objections with the incompleteness argument, that is, by pointing out that it has never been realized in its pure form and that its shortcomings can always be traced to exogenous influences. Such a procedure is not justifiable from an ethical standpoint. A social theory must adjust itself to reality and take its historical locus and historical

123 JÜNGER (1952) p. 37.

conditions of realization into account. A theory that presents a superior model but can never be realized due to exogenous influences or non-producible preconditions is a bad utopia and remains in the „precociousness of the obligation" (Hegel).

As a theory of society, capitalism cannot suffice because it is essentially an economic theory of production, exchange, and coordination. As an economic theory, it must neglect essential aspects of social action and political integration. In this work, the omission of problems prevailing in many approaches to the theory of capitalism has been exhibited above all in the assumption of given preferences in neoclassical economic theory but also in the limits to the principle of coordination by markets. In both cases, the cause is to be found in an exaggerated methodological individualism and subjectivism, which assumes one can neglect the social mediation of one's preferences and the obligation to have reasonable preferences for the sake of individual freedom of choice.

By making freedom and efficiency the sole guiding values in its ideal of coordination, the theory of capitalism evades the problem of comparing goods and gains its most impressive comprehensiveness in general equilibrium analysis. The problem of weighing goods against one another cannot be avoided, however, as the problem of weighing sacrifices of liberty for the sake of increases in efficiency and vice versa demonstrate. This was already seen with the necessity of weighing efficiency against freedom in the selection of allocation mechanisms.

The necessity of comparing goods also turns up in the preference formation of the individual, which cannot be regarded as a black box from which factual preferences pass into making a choice. The mechanism model of choice and decision making that interprets the actual choices and revealed preferences as the effects of inexaminable forces, the inner preferences, successfully avoids illuminating the „black box" that is the preference formation for the revealed preference approach. But precisely the preference formation gains significance when with increasing social wealth and material satiation the production problem becomes less urgent.[124]

124 The excellent books by HIRSCH (1976) and SCITOVSKY (1976) demonstrate the insufficiency of traditional microeconomic theory upon the analysis of preference formation and the satisfaction of needs and, thereby, show the limits of its suitability for policy recommendations, especially for those concerning economic growth policy. - The intermediate structure of the culture of the economy, *der Zwischenbau der Kultur* as Gustav Schmoller called it, gains more and more importance in relation to the substructure of eco-

2. Freedom and Freedom of Choice

This satiation also demonstrates the limitation of a concept of freedom that understands freedom merely as freedom of choice between the greatest number of possibilities. Quite often it turns out that a growing supply of goods does not cause that one feels freer when one has greater possibilities of choice among goods. Rather, such situations of an oversupply of alternatives demonstrate that the marginal utility of freedom of choice is also decreasing. Too many possibilities is just as unpleasant as too few alternatives, not merely due to overburdening of decision-making and problems of cognitive dissonance after consumption decisions but because this concept of freedom systematically touches on only one part of what human freedom is.

Philosophical criticism caricatured the understanding of freedom as freedom of choice in the High Middle Ages with Buridan's ass, which starves because it cannot decide between two hay stacks of equal size, and, in Modern Times since Kant, as the freedom of a turnspit. Moreover, in the mechanism model of general equilibrium theory there can be no real freedom. The agent is determined by his preferences and mechanically adapts himself by his consumption and production decisions to the prevalent market conditions.

Against this conception of economic freedom it must be stressed that freedom primarily means the ability to act according to self-chosen purposes, and that the choice of these purposes or the formation of preferences must be understood as a self-choice, that is, a decision about one's own being and personality.[125] It follows that the concentration of economic theory upon consumption and consumption decision as the purpose of human and, in particular, economic action is one-sided because the personality forms itself essentially in its action and creative occupation not in its consumption. This implies that an examination of the extent to which the constitution of an economy admits free action and self-realization cannot be based merely on freedom of consumption.

The indubitable superiority of the market as a means of coordination manifests itself here in that it permits individual pursuit of goals and self-respon-

---

nomic exchange andto the superstructure of law. Cf. for Schmoller and the Historical School of Economics KOSLOWSKI (1991). The economy is itself part of the society's culture as is shown in KOSLOWSKI (1989).

125 See F.W.J. SCHELLING: *Philosophische Untersuchungen über das Wesen der menschlichen Freiheit* (1809), ed. W. Schulz, Frankfurt 1975, p. 77; KRINGS (1980, pp. 15-39).

sible action to a greater extent than all other forms of coordination, in that *the market permits not only freedom of consumption, but also freedom of action and production.*

3. The Necessity of a Social and Ethical Framework for Capitalism

It also turns out that the maximization of profit and utility as the economic motive, and the free disposition over private property assume a characteristic abstractness when they lay claim to unlimited social validity. The maximization of profit and utility can only be admitted as motives under constraints; otherwise they reduce the wealth of human motivation to abstractions of rationality and ignore the social embeddedness of the pursuit of goals. The same is true of rights of disposition over private property.[126] The coordination of individual actions in capitalism must occur within a social framework which the conditions of this coordination—private property, maximization of profit and utility, and the market system—do not adequately determine, but rather presuppose.[127]

The limits of capitalism as a social theory are that the coordination ideal does not comprise the whole of a society's guiding ideas, that the medium is not the message, and that the form of economic action does not fully comprise the substance of one's social action. As a social theory, capitalism is *materially underdetermined* and incomplete. It must be complemented by a comprehensive social-philosophical theory concerning the framework within which capitalism can activate its advantage as a method of coordination, by a theory of the social genesis and normative justification of preference formation (social psychology and ethics), by a theory of the social institutions of which this framework consists (family, churches and religious communities, state), and by a theory of political compensation for capitalism failure (market failure, the limits of subjectivism, and consideration of substantial life interests).

---

126 See KOSLOWSKI (1995) for an application in the ethics of banks and financial markets.

127 NELL-BREUNING aptly criticizes the theory of the market economy in that „more and more, problems are pushed into the so-called data wreath, the framework, and consequently the data wreath ultimately becomes that which is most interesting" (1955, p. 111).

The necessity of such a framework becomes evident in the dialectic proper to the three structural characteristics of capitalism. In all three characteristics a change-over from quantity into quality and from form into content is observable. The unlimited accumulation of private property leads, from a certain point of mastery of a market to a qualitative jump and to a problem of power. Unlimited pursuit of profits and benefits leads to a change-over into greed, miserliness and a loss in the wealth of human purposes. The coordination of production and the assignment of social status exclusively by way of market success, that is, successful anticipation of demand and willingness to pay, leads to an exaggerated subjectivism and the neglect of more substantial purposes.

The *form* of coordination by way of property, maximization of utility and profit, and the market cannot be the *content* of the social order and individual action, no more so than this form can be abandoned if freedom and efficiency in the economy are to be secured. The theory of capitalism requires a complement from social philosophy and a reminder that *reasonable preferences* must enter into the coordination. It also needs the reminder that capitalism lives from an ethos of freedom and work, which as a form of economic coordination it alone cannot bring forth and preserve. An ideal of coordination alone cannot do justice to our need for substantial life forms, just as, on the other hand, our need for the recognition of our subjectivity and freedom, in the economy as well, requires the capitalist form of coordination.

References

AEGIDIUS DE LESSINIA: *De usuris in communi* (1276/85), in: THOMAS AQUINAS: *Opera Omnia* (ed. Fretté), Paris 1875, Vol. 28.

ALBACH, H.: „Vertrauen in der ökonomischen Theorie," *Zeitschrift fur die gesamte Staatswissenschaft*, 136 (1980), no. 1; 3.

ALCHIAN, A. A.: „Uncertainty, Evolution, and Economic Theory", in: A. A. ALCHIAN (ed.): *Economic Forces at Work*, Indianapolis (Liberty Press) 1977, pp. 15-36.

ALCHIAN, A. A. and H. DEMSETZ: „Production, Information Costs and Economic Organization", in: A. A. ALCHIAN (ed.): *Economic Forces at Work*, Indianapolis (Liberty Press) 1977, pp. 73-110.

ARISTOTLE: *Metaphysics*, ed. W. D. Ross, 2 Vols., Oxford (Oxford University Press) 1924.

ARISTOTLE: *Nicomachean Ethics*, ed. I. Bywater, Oxford (Oxford University Press) 1894.

ARISTOTLE: *Politics*, ed. W. D. Ross, 2 Vols., Oxford (Oxford University Press) 1957.

ARROW, K. J.: „Public and Private Values", in: S. HOOK (ed.): *Human Values and Economic Policy*, New York (New York University Press) 1967.

ARROW, K. J.: „Political and Economic Evaluation of Social Effects and Externalities", in: M. D. INTRILIGATOR (ed.): *Frontiers of Quantitative Economics*, Amsterdam (North Holland) 1971.

BANFIELD, E. C.: *The Moral Basis of a Backward Society*. Glencoe, Ill. (The Free Press) 1958.

BAUMOL, W. J.: „Business Responsibility and Economic Behavior", in: E. S. PHELPS (ed.): *Altruism, Morality and Economic Theory*, New York (Sage Foundation) 1975.

BECKER, GARY: „A Theory of Social Interactions", *Journal of Political Economy*, 82 (1974).

BELL, D.: *The Cultural Contradictions of Capitalism*, New York (Basic Books) 1976.

BLUMENBERG, H.: „Selbsterhaltung und Beharrung. Zur Konstitution der neuzeitlichen Rationalität", in: H. EBELING (ed.): *Subjektivität und Selbsterhaltung. Beiträge zur Diagnose der Moderne*, Frankfurt (Suhrkamp) 1976, pp. 144-207.

BÖHM-BAWERK, E. V.: *Kapital and Kapitalzins*, Vol. 1.: *Geschichte und Kritik der Kapitalzinstheorien*, Innsbruck 1884.

REFERENCES

BONAR, J.: *Philosophy and Political Economy in Some of Their Historical Relations* (1893). Reprint London (S. Sonnenschein) 1909.

BOULDING, K. E.: „The Basis of Value Judgements in Economics", in: S. HOOK (ed.): *Human Values and Economic Policy*, New York (New York University Press) 1967, pp. 55-72.

BOULDING, K. E.: „Economics as a Moral Science", *American Economic Review*, 59 (1969).

BRENTANO, LUJO: *Ethik und Volkswirtschaft in der Geschichte*, München 1902.

BRUNNER, OTTO: „Die alteuropäische Ökonomik", *Zeitschrift für Nationalökonomie*, 13 (1950).

BUCHANAN, JAMES M. (1954a): „Social Choice, Democracy, and Free Markets", *Joumal of Political Economy*, 62 (1954), pp. 114-123.

BUCHANAN, JAMES M. (1954b): „Individual Choice in Voting an the Market", *Joumal of Political Economy*, 62 (1954), pp. 334-343.

BUCHANAN, JAMES M.: „What Should Economists Do?", *Southern Economic Journal*, 30 (1964), pp. 213-222.

BUCHANAN, JAMES M.: „Ethical Rules, Expected Values, and Large Numbers", *Ethics*, 76 (1965), pp. 1-13.

BUCHANAN, JAMES M.: „Is Economics the Science of Choice?", in: E. STREISSLER (ed.): *Roads to Freedom: Essays in Honour of Friedrich A. von Hayek*, New York (A.M. Kelly) 1969.

BUCHANAN, JAMES M.: „Markets, States, and the Extent of Morals", *American Economic Review*, 68 (1978), pp. 364-368.

BUCHANAN, JAMES M. and G. TULLOCK: *The Calculus of Consent*, 5th ed., Ann Arbor (University of Michigan Press) 1974.

CHURCHMAN, C. W.: *Prediction and Optimal Decision, Philosophical Issues of a Science of Values*, Englewood Cliffs, N.J. (Prentice-Hall) 1961.

COASE, R. H.: „The Problem of Social Cost", *Journal of Law & Economics*, 3 (1960), pp. 1-44.

DÖLLINGER, I. V., and FR. H. REUSCH: *Geschichte der Moralstreitigkeiten in der römisch-katholischen Kirche*, Nördlingen (C.H. Beck) 1889.

DOI, TAKEO: *The Anatomy of Dependence*, Tokyo, New York, London (Kodanska) 1973. Japanese Original: *Amae no Kozo*, Tokyo (Kobundo) 1971.

DUMONT, L.: *From Mandeville to Marx: The Genesis and Triumph of Economic Ideology*, Chicago (University of Chicago Press) 1977.

DUPUY, J. P.: „L'economie de la morale, ou la morale de l'economie", *Revue d'Economie Politique*, 88 (1978), pp. 404-439.

EBELING, H. (ed.): *Subjektivität und Selbsterhaltung. Beiträge zur Diagnose der Moderne*, Frankfurt (Suhrkamp) 1976.

EPICTETUS, *Dissertationes*, ed. JOH. SCHWEIGHÄUSER, Leipzig 1799, Reprint Hildesheim/New York (Olms) 1977.

FICHTE, J. G.: *Der geschlossene Handelsstaat* (1800), in: J.G. FICHTE: *Werke*, ed. by I. H. Fichte, Reprint Berlin (de Gruyter) 1971, Vol. 3.

FOSTER, G. M.: „Peasant Society and the Image of Limited Good", *American Anthropologist*, 67 (1965), pp. 293ff.

FREYER, H.: *Die Bewertung der Wirtschaft im philosophischen Denken des 19. Jahrhunderts* (1921). Reprint Hildescheim (Olms) 1966.

FRITZ, K. V.: *The Theory of the Mixed Constitution in Antiquity*, New York (Columbia University Press) 1954.

GILLESPIE, M.: Review of P. Koslowski: Ethik des Kapitalismus, *Ethics. International Journal of Social Philosophy*, 94 (Chicago, Oct 1983), Nr. 3.

GROSSMANN, H.: „Die gesellschaftlichen Grundlagen der mechanistischen Philosophie und die Manufaktur", *Zeitschrift für Sozialforschung*, 4 (1935), pp. 161-231, reprint Munich (Kösel) 1970.

HALLOWELL, A. I.: „Ojibwa ontology, behavior, and world view, in: S. DIAMOND (ed.): *Culture in History. Essays in Honor of Paul Radin*, New York (Columbia University Press) 1960, pp. 19-52.

HARTWELL, R. M.: „The Origins of Capitalism: A Methodological Essay", in: S. PEJOVICH (ed.): *Philosophical and Economic Foundations of Capitalism*, Lexington (D. C. Heath) 1983, pp. 11-24.

HAYEK, F. A. VON: *Der Wettbewerb als Entdeckungsverfahren*, Kiel (Institut für Weltwirtschaft) 1968.

HAYEK, F. A. VON: 1969. *Freiburger Studien*, Walter Eucken Institut (ed.), Wirtschaftswissenschaftliche und wirtschaftsrechtliche Untersuchungen 5, Tübingen (J.C.B. Mohr [Paul Siebeck]) 1969.

HAYEK, F. A. VON: *Drei Vorlesungen über Demokratie, Gerechtigkeit und Sozialismus*, Walter Eucken Institut (Hrsg.), Vorträge and Aufsätze 63, Tübingen (J.C.B. Mohr [Paul Siebeck]) 1977.

HEGEL, GEORG WILHELM FRIEDRICH: *Philosophy of Right*, translated with notes by T. M. Knox, Oxford 1942. German Original: *Grundlinien der Philosophie des Rechts* (1821), Frankfurt a. M. (Suhrkamp) 1970.

HENRICH, D.: „Die Grundstruktur der modernen Philosophie", in: H. EBELING (ed.): *Subjektivität und Selbsterhaltung. Beiträge zur Diagnose der Moderne*, Frankfurt (Suhrkamp) 1976, pp. 97-144.

HIRSCH, F.: *Social Limits to Growth*, Cambridge, Mass. (Harvard University Press) 1976.

HIRSCHMAN, A. O.: *Exit, Voice, and Loyalty: Responses to Decline in Firms, Organizations, and States*, Cambridge, Mass. (Harvard University Press) 1970.

HIRSCHMAN, A. O.: *The Passions and the Interests: Political Arguments for Capitalism before its Triumph*, Princeton, N. J. (Princeton University Press) 1977.

HIRSHLEIFER, J.: „Competition, Cooperation, and Conflict in Economics and Biology", *American Economic Review*, 68 (1978), pp. 238ff.

HOBBES, TH.: *Leviathan* (1651), ed. C. B. Macpherson, London (Penguin) 1968.

HÖFFNER, J.: *Statik und Dynamik in der scholastischen Wirtschaftsethik*, Köln (Westdeutscher Verlag) 1955.

REFERENCES

HOOK, S. (ed.): *Human Values and Economic Policy*, New York (New York University Press) 1967.

HUTCHISON, T. W.: „Notes on the Effects of Economic Ideas on Policy: The Example of the German Social Market Economy", *Zeitschrift für die gesamte Staatswissenschaft*, 135 (1979) pp. 433ff.

JOHNSON, P.: „Is there a Moral Basis for Capitalism? Dissenting Thoughts in a Collectivist Age", *Encounter* (October 1979), pp. 15-25.

JÜNGER, E.: *Der Waldgang*, Frankfurt (Klostermann) 1952.

KANT, I.: *Critique of Practical Reason*, London (Longmans, Green & Co.) 1898. German Original: *Kritik der praktischen Vernunft* (1788), Akademie-Ausgabe der Preußischen Akademie der Wissenschaften, Berlin 1902ff., reprint Berlin (de Gruyter) 1968.

KNIGHT, F. H. (1935a): „The Ethics of Competition",

KNIGHT, F. H. (1935b): „Ethics and the Economic Interpretation",

KNIGHT, F. H. (1935c): „Value and Price",

KNIGHT, F. H. (1935d): „Statics and Dynamics. Some Queries Regarding the Mechanical Analogy in Economics",

KNIGHT, F. H. (1935e): „The Limitations of Scientific Method in Economics", all papers in: F. H. KNIGHT: *The Ethics of Competition and Other Essays* (1935). Reprint Freeport, N.Y. (Books for Libraries Press) 1969.

KNIGHT, F. H. (1947a): „Freedom as Fact and Criterion",

KNIGHT, F. H. (1947b): „The Sickness of Liberal Society", both papers in: F. H. KNIGHT: *Freedom and Reform*, New York (Harper) 1947.

KNIGHT, F. H.: „Abstract Economics as Absolute Ethics", *Ethics*, 76 (1966), pp. 163-177.

KNOLL, A. M.: *Zins und Gnade. Zur Zins-und Gnadenkontroverse der Dominikaner and Jesuiten, Lutheraner and Calviner*, Neuwied (Luchterhand) 1967.

KOSLOWSKI, PETER (1979a): „Haus and Geld. Zur aristotelischen Unterscheidung von Politik, Ökonomik and Chrematistik", *Philosophisches Jahrbuch*, 86 (1979), pp. 60-83.

KOSLOWSKI, P. (1979b): *Zum Verhältnis von Polis and Oikos bei Aristoteles, Politik und Ökonomie bei Aristoteles*, 2nd ed. 1979, 3rd ed. Tübingen (J.C.B. Mohr [Paul Siebeck]) 1993 (1st ed. 1976).

KOSLOWSKI, P. : *Gesellschaft und Staat. Ein unvermeidlicher Dualismus*, Stuttgart (Klett-Cotta) 1982.

KOSLOWSKI, P. (1983 a): „Markt- *und* Demokratieversagen? Grenzen individualistischer gesellschaftlicher Entscheidungssysteme", *Politische Vierteljahresschrift*, 24 (1983), pp. 166-187.

KOSLOWSKI, P. (1983 b): „Mechanische und organistische Analogien in der Wirtschaftswissenschaft - eine verfehlte Alternative", *Kyklos. Internationale Zeitschrift für Sozialwissenschaften*, 36 (1983), pp. 308-312.

KOSLOWSKI, P.: Article „Nebenwirkungen", in: *Historisches Wörterbuch der Philosophie*, Vol. 6, ed. by J. RITTER and K. GRÜNDER, Basel (Schwabe) 1984, pp. 659-662.

KOSLOWSKI, P.: „Market and Democracy as Discourses. Limits to Discoursive Social Coordination", in: P. KOSLOWSKI (eds.): *Individual Liberty and Democratic Decision-Making. The Ethics, Economics, and Politics of Democracy*, Tübingen (J.C.B. Mohr [P. Siebeck]) 1987, pp. 58-92.

KOSLOWSKI, P.: *Prinzipien der Ethischen Ökonomie. Grundlegung der Ethik und der auf die Ökonomie bezogenen Ethik*, Tübingen (J.C.B. Mohr [Paul Siebeck]) 1988, 2nd ed. 1994.

KOSLOWSKI, P.: *Wirtschaft als Kultur. Wirtschaftsethik und Wirtschaftskultur in der Postmoderne*, Wien (Edition Passagen) 1989.

KOSLOWSKI, P.: *Gesellschaftliche Koordination. Eine ontologische und kulturwissenschaftliche Theorie der Marktwirtschaft*, Tübingen (J.C.B. Mohr [Paul Siebeck]) 1991.

KOSLOWSKI, P.: *Die Ordnung der Wirtschaft. Studien zur Praktischen Philosophie und Politischen Ökonomie*, Tübingen (J.C.B. Mohr [Paul Siebeck]) 1994.

KOSLOWSKI, P.: „The Ethics of Banking. On the Ethical Economy of the Credit and Capital Market, of Speculation and Insider Trading in the German Experience", in: A. ARGANDOÑA (ed.): *The Ethical Dimension of Financial Institutions and Markets*, Berlin, Heidelberg, New York, Tokyo (Springer) 1995, pp. 180-232 (= Studies in Economic Ethics and Philosophy, Vol. 8).

KOSLOWSKI, P.: „Ecology and Ethics in the Economy", in: F. N. BRADY (ed.): *Ethical Universals in International Business*, Berlin, Heidelberg, New York, Tokyo (Springer) 1996 (= Studies in Economic Ethics and Philosophy, Vol. 9).

KOSLOWSKI, P. (ed.): *Ethics in Economics, Business, and Economic Policy*, Berlin, Heidelberg, New York, Tokyo (Springer) 1992 (= Studies in Economic Ethics and Philosophy, Vol. 1).

KOSLOWSKI, P. (ed.): *The Theory of Ethical Economy in the Historical School. Wilhelm Roscher, Lorenz von Stein, Gustav Schmoller, Wilhelm Dilthey and Contemporary Theory*, Berlin, Heidelberg, New York, Tokyo (Springer) 1995 (= Studies in Economic Ethics and Philosophy, Vol. 7).

KOSLOWSKI, PETER and YUICHI SHIONOYA (eds.): *The Good and the Economical*, Berlin, Heidelberg, New York, Tokyo (Springer) 1993 (= Studies in Economic Ethics and Philosophy, Vol. 4).

KRINGS, H.: *System und Freiheit. Gesammelte Aufsätze*, Freiburg (Alber) 1980.

KRINGS, H.: *Ordo. Philosophisch-historische Grundlegung einer abendländischen Idee*, Hamburg (Meiner) 1982, 1st ed. 1941.

KROMPHARDT, J.: *Konzeptionen und Analysen des Kapitalismus*, Göttingen (Vandenhoeck) 1980.

LEIBNIZ, G. W.: *Die Theodizee* (1710), tr. by A. Buchenau, Hamburg (Meiner) 1968.

REFERENCES

LUHMANN, N.: *Zweckbegriff und Systemrationalität. Über die Funktion von Zwecken in sozialen Systemen*, 2d ed. Frankfurt (Suhrkamp) 1977.

MACPHERSON, C. B.: *The Political Theory of Possessive Individualism: Hobbes to Locke*, London (Oxford University Press) 1962.

MACRAE, N.: „Für eine Welt individueller Lebensstile", in: R. LÖW, P. KOSLOWSKI, P. KREUZER (eds.): *Fortschritt ohne Maß? Eine Ortsbestimmung der wissenschaftlich-technischen Zivilisation*, München (Piper) 1981, pp. 213-233.

MANDEVILLE, B. DE : *The Fable of the Bees: or, Private Vices, Publick Benefits* (1714-1729), ed. F. B. Kaye, Oxford 1924.

MARSHAK, J.: „Economic Planning and the Cost of Thinking", in: J. MARSHAK: *Economic Information, Decision and Prediction*, Dordrecht (Reidel) 1974, Vol. 2, pp. 193-199.

MARX, K.: „Manifest der kommunistischen Partei (1848)", in: MARX/ENGELS, *Werke*, Berlin (Dietz) 1977, Vol. 4.

MIRABEAU, VICTOR MARQUIS DE: *Letter of March 31st, 1770, to the Markgraf Carl Friedrich of Baden*, ed. by Carl Knies, Heidelberg (Winter) 1892, Vol. I.

MISES, L. VON: *Human Action*, New Haven (Yale University Press) 1949.

MOLINA, LUIS DE: *De justitia et jure*, Moguntiae 1602.

NELL-BREUNING, O. V.: „Neoliberalismus und Katholische Soziallehre", in: P. M. BOARMANN (ed.): *Der Christ und die soziale Marktwirtschaft*, Stuttgart (Kohlhammer) 1955.

NELL-BREUNING, O. V.: *Kapitalismus-kritisch betrachtet*, Freiburg (Herder) 1974.

NIDER, JOANNES (ca. 1380-1438): *Tractatus de contractibus mercatorum*, Paris circa 1495 (Munich: Bayerische Staatsbibliothek, Sign. Nr. 4 1341). English translation: JOHANNES NIDER: *On the Contracts of Merchants*, translated by Charles H. Reeves, edited by Ronald B. Shuman, Norman (University of Oklahoma Press) 1966.

NORMAN, EDWARD R.: „Denigration of Capitalism", in: M. NOVACK (ed.): *The Denignation Capitalism: Six Points of View*, Washington, D.C. (American Enterprise Institute) 1979, pp. 7-23.

NOVACK, M. (ed.): *The Denigration of Capitalism: Six Points of View*, Washington, D.C. (American Enterprise Institute) 1979.

NOZICK, R.: *Anarchy, State, and Utopia*. Oxford (Basil Blackwell) 1974.

PASSOW, R.: *Kapitalismus. Eine begrifflich-terminologische Studie*, Jena (Fischer) 1927.

PHELPS, E. S. (ed.): *Altruism, Morality and Economic Theory*, New York (Sage Foundation) 1975.

PLATO: *Gorgias*, revised text with introduction, commentary, and appendix on Socrates, Callicles, and Nietzsche by E. R. Dodds, Oxford (Oxford University Press) 1959.

PLATO: *Plato's Examination of Pleasure*, translation of the *Philebus*, with introduction and commentary by R. Hackforth, Cambridge 1945.

POLANYI, K.: *Primitive, Archaic and Modern Economies*, Boston (Beacon Press) 1971.

POSNER, R. A.: *Economic Analysis of Law*, Boston (Little, Brown) 1977.

RÖPKE, JOCHEN: *Primitive Wirtschaft, Kulturwandel und die Diffusion von Neuerungen*, Tübingen (J.C.B. Mohr [Paul Siebeck]) 1970.

RÖPKE, WILHELM: *Civitas Humana: Grundfragen der Gesellschafts-und Wirtschaftsreform*, Erlenback-Zürich (Rentsch) 1949.

ROSCHER, W.: *Geschichte der Nationalökonomik in Deutschland*, München (Oldenbourg) 1874.

ROWLEY, C. K., and A. T. PEACOCK: *Welfare Economics: A Liberal Restatement*, London (Martin Robertson) 1975.

RÜSTOW, A.: *Das Versagen des Wirtschaftsliberalismus als religions-geschichtliches Problem*, Zürich and New York (Europa Verlag) 1945.

SAHLINS, M. D.: „Exchange-Value and the Diplomacy of Primitive Trade", in: *Essays in Economic Anthropology: Dedicated to the Memory of Karl Polanyi*, Seattle (University of Washington Press) 1965.

SALIN, EDGAR: *Politische Ökonomie*, Tübingen (J.C.B. Mohr [Paul Siebeck]) 1967.

SAUERMANN, HEINZ (ed.): *Bargaining Behavior*, Tübingen (J.C.B. Mohr [Paul Siebeck]) 1978.

SCHALL, J. V.: „Religion and the Demise of Capitalism", in: M. NOVACK (ed.): *The Denigration of Capitalism: Six Points of View*, Washington, D.C. (American Enterprise Institute) 1979, pp. 32-38.

SCHELER, MAX: *Der Formalismus in der Ethik und die materiale Wertethik*, 5th ed. Bern (Francke) 1966.

SCHELLING, F.W.J.: *Philosophische Untersuchungen über das Wesen der menschlichen Freiheit* (1809), ed. W. Schulz, Frankfurt 1975.

SCHLOSSER, JOHANN GEORG: *Über das neue französische System der Policeyfreyheit, insbesondere in der Aufhebung der Zünfte* (About the new French system of police freedom, especially in the abolition of the guilds), *Kleinere Schriften* 1. Theil Basel 1787, Reprint New York/London (Johnso) 1972.

SCHUMACHER, E. F.: *Small Is Beautiful*, London (Blond and Briggs) 1973.

SCHUMPETER, J. A.: *History of Economic Analysis*. 5th ed. London (Allen & Unwin) 1955.

SCITOVSKY, T.: *The Joyless Economy*. London (Oxford University Press) 1976.

SEN, A.: „Isolation, Assurance and the Social Rate of Discount", *Ouarterly Joumal of Economics*, 81 (1967), pp. 112-124.

SEN, A.: *On Economic Inequality*, Oxford (Clarendon Press) 1973.

SHACKLE, G.L.S.: *Epistemics and Economics: A Critique of Economic Doctrines*, Cambridge (Cambridge University Press) 1972.

SIMON, H.A.: „Rationality as Process and as Product of Thought", *American Economic Review*, 68 (1978), pp. 1-15.

REFERENCES

SINGER, K.: „Oikonomia: An Inquiry into Beginnings of Economic Thought and Language", *Kyklos*, 11 (1958), pp. 29-57.

SOMBART, W.: *Der moderne Kapitalismus*, Leipzig (Duncker & Humblot) 1902ff.

SOMBART, W.: *Luxus und Kapitalismus* (1912). Reprinted in the version of the 2nd edition of 1922 under the title: *Liebe, Luxus und Kapitalismus. Über die Entstehung der modernen Welt aus dem Geist der Verschwendung*, Berlin (Wagenbach) 1984.

SOMBART, W.: *Die Zukunft des Kapitalismus*, Berlin (Buchholz & Weißwange) 1937.

SOMBART, W.: *Weltanschauung, Wissenschaft und Wirtschaft*, Berlin (Buchholz and Weisswange) 1938.

SPAEMANN, R: *Reflexion und Spontanität. Studien über Fenélon*, Stuttgart (Kohlhammer) 1963.

SPAEMANN, R.: „Nebenwirkungen als moralisches Problem", in: R. SPAEMANN: *Zur Kritik der politischen Utopie*, Stuttgart (Klett) 1977.

SPAEMANN, R.: „Christentum und Kernkraft. Ethische Aspekte der Energiepolitik", *Die politische Meinung*, 192 (Osnabrück 1980).

SPENCER, H.: *The Man Versus the State*, Indianapolis (Liberty Classics) 1982. Ffirst ed. 1884.

STAHL, FRIEDRICH JULIUS: *Die Philosophie des Rechts*, 5 Vols., 1878, reprint Darmstadt 1963.

STEGMÜLLER, W.: *Probleme und Resultate der Wissenschaftstheorie und Analytischen Philosophie. Personelle und statistische Wahrscheinlichkeit. 1. Halbbd.: Personelle Wahrscheinlichkeit und rationale Entscheidung*, Berlin (Springer) 1973.

STRAUSS, L.: *Natural Right and History*, Chicago (University of Chicago Press) 1953.

STREISSLER, E. and CH. WATRIN (eds.): *Zur Theorie marktwirtschaftlicher Ordnungen*, Tübingen (J. C. B. Mohr [Paul Siebeck]) 1980.

TAWNEY, R. H.: *Religion and the Rise of Capitalism*, 2d ed. London (John Murray) 1937.

THOMAS AQUINAS: *Summa theologiae*, Vols. IV-XII of the Leonine edition, Rome 1918-1930, reprinted Turin 1934. Tranlated by the English Dominican Fathers as *The Summa Theologica*, 22 Vols., London 1912-1936.

TROELTSCH, E.: *Die Soziallehren der christlichen Kirchen and Gruppen*, Tübingen (J.C.B. Mohr [Paul Siebeck]) 1923.

VINER, J.: *The Role of Providence in the Social Order: An Essay in Intellectual History*, Philadelphia (American Philosophical Society) 1972.

WAGNER, F.: *Das Bild der frühen Ökonomik*, Salzburg (Stifterbibliothek) 1969.

WEBER, MAX: *Wirtschaft und Gesellschaft*, 5th ed. Tübingen (J.C.B. Mohr [Paul Siebeck]) 1972.

WEBER, MAX: *Die protestantische Ethik und der Geist des Kapitalismus*, Gütersloh (Mohn) 1979.

WEGEHENKEL, L. (ed.).: *Marktwirtschaft und Umwelt,* Walter Eucken Institut, Wirtschaftswissenschaftliche und wirtschaftsrechtliche Untersuchungen 17, Tübingen (J. C. B. Mohr [Paul Siebeck]) 1981.
WEISSKOPF, W. A. : *Alienation and Economics,* New York (Dutton) 1971.
WILLIAMSON, O. E.: „Firms and Markets", in: S. WEINTRAUB (ed.): *Modern Economic Thought,* Philadelphia (University of Pennsylvania Press) 1977.
WILLIAMSON, O. E.: „The Modern Corporation: Origins, Evolution, Attributes", *Journal of Economic Literature,* 19 (1981), pp. 1537-1570.

The Morality of Capitalism Comment on Peter Koslowski, The Ethics of Capitalism

by JAMES M. BUCHANAN

My comment is in two sections. In Section One, I shall refer specifically to Koslowski's paper. In Section Two, I shall offer more general remarks on „The Morality of Capitalism", remarks stimulated by reading Koslowski's paper as well as other sources, and reflecting an effort on my part to incorporate several aspects of the topic into my own perspective.

Section One

Comment on Peter Koslowski

Peter Koslowski deserves commendations on his response to the challenge put to him in the invitation to prepare a paper on the morality of capitalism. As he suggests, his task was surely the most delicate and also the most difficult of the conference. Considered in the large, his paper must be rated „Excellent." He has raised the relevant questions, given us a feel for the opposing views on these, provided the English-language provincials with an exposure to much relevant classical and European literature, for the most part outside economics, and, finally, offered several suggestions toward the construction of a genuinely defensible morality of capitalism.

But excellent papers deserve criticism, and what I want to do now is to suggest that Koslowski's discussion may be faulted in several particulars. Let me turn first to page 35, on which I think there is an error, at least in my interpretation of the argument. At the top of this page, Koslowski argues, commendably, that a theory of the formation of preferences is logically prior to a theory of the allocation of means to meet those preferences. So far, so good. But Koslowski then goes on to challenge the moral status of the

Pareto norm, even if we disregard the theory problem of the initial or pre-trade distribution of endowments.

If a distribution of endowments is given, *and if preferences are what they are,* then there is positive moral content in the Pareto norm for the elementary reason that the Pareto optimum describes the end-state to which freedom of individual choice will lead in the absence of arbitrary constraints. I do not understand the sentence in the paper which reads: „In as far as each movement on the contract curve presupposes a certain indulgence and the absence of jealousy or envy (BOULDING, 1969), the Pareto optimum implies a moral minimum.“ This statement seems to me to be confused and seriously misleading.

Indulgence, jealousy, envy, altruism -- what have you -- these can all be incorporated into preferences, and from these preferences indifference maps may be constructed for pedogogic purposes. Whether or not these elements of character are or are not present has nothing to do with the moral significance of the Pareto norm. This moral significance lies *within* the theory of allocation and is *not relevant* to the theory of preference formation, to use Koslowski's useful two-part classification here. To be sure, we should all like a world in which envy is absent better than one in which envy is omnipresent, but this amounts to saying that we like one set of preferences better than another. But even in the latter world, the Pareto norm is, in effect, an economic translation of the more basic norm of allowing individuals to express their own preferences, whatever these might be. I shall discuss this point indirectly again in Section Two of this comment.

I should enter what may seem a trivial objection to Koslowski's discussion at the top of page 43. Following Knight's attribution of rewards to effort, inheritance, and luck, Koslowski suggests that the third of these, luck, is „incommensurable with justice“. From this it follows that, at least, luck-based reward can be labelled „not immoral“. In my view, the emphasis here should be on the latter statement. What is important in the establishment of „legitimacy“ for a set of social institutions, is not so much any positive evaluation of these institutions in terms of criteria of morality or justice, but, instead, the *absence* of negative evaluation which describes the „rules of the game“ as unfair or unjust. Being unlucky in a game of fair rules is not likely to be serious cause for social discontent. Constructive design should place emphasis on insuring that the „game" be seen to be „fair".[1]

1 I have elaborated on some of these and related points in two lectures delivered December 1979, and published as J. M. BUCHANAN: „Fairness, Hope,

I share Koslowski's judgment (expressed at the bottom of page 44) that a purely deontological or *process* theory of justice makes little sense. Process must be at least in part evaluated in terms of predicted consequences. At the end of the paragraph (top of page 44), however, Koslowski mentions „the primary distribution which results from the market in capitalism." I do not know what these words mean. The „market" cannot exist independently of a legal-political framework and there are many variations among legal-political institutions all of which may allow for the operation of something we might call „a market" and each one of which might generate a differing probability distribution of imputations of final product, ***even if we postulate a given pre-trade distribution of endowments among persons.*** Any final distributive result that might be observed emerges from several separate elements. It is one particular distribution from among a probability distribution of distributions, even given a set of initial endowments, a specific legal-political framework, and the operation of the market. Koslowski comes dangerously close to slipping into a very common error here of attributing to „the market" all objectionable features about final distributions, which may be more properly attributed to personal pique at being unlucky, to badly-designed laws, and, most importantly, to objectionable features in the distribution of initial endowments.

Let me shift to page 47 and challenge Koslowski's statement: „For the market to be moral and the point it chooses on the production possibility frontier to be reasonable, would require that the effective demand, that is, the preferences of consumers were moral and their knowledge perfect." This statement seems to me to reflect a failure to appreciate the significance of the distinction between the two levels of discourse that Koslowski introduced earlier in the paper, the theory of preference formation and the theory of allocation. I shall indirectly discuss this distinction again in Section Two. I shall finish Section One of my comment by four citations, presented without discussion from Koslowski's paper. Each one of these will bear some relevance to the more general theme I want to develop in Section Two:

(Page 48):

„When the religious formation of preferences is abandoned, other forces for molding preferences, other Gods and idols appear."

---

and Justice", in: ROGER SKURSKI (ed.): *New Directions in Economic Justice*, South Bend, Indiana (University of Notre Dame Press) 1983, pp. 53-89.

(Page 49):

„That capitalism, despite its successes, has such a bad reputation is to be regarded as a result of the not-understood problem of the non-finalization of the economy."

(Page 51):

„That which the individual and society takes to be preferred can only be reached by the market through a compromise between what the individual takes to be important and what all others take to be important."

(Page 51):

„The terror of social determination of goals by *one* group is the alternative to competition in advanced societies because, according to historical experience, only through terror can a society which has experienced freedom be sworn to a particular goal."

Section Two

The Morality of Capitalism

In this part of my comment, I shall introduce four separate elements to construct an argument that I hope will both clarify some of the issues raised under the rubric „the morality of capitalism" and offer some suggestions concerning the potential for deriving an acceptable morality. The separate and apparently disparate elements are:

1. Koslowski's stress on the conflict between the autonomy of the individual in capitalistic economies and what he calls „value rationality" in society;
2. Michael Polanyi's diagnosis of the modern historical experience as characterized by „moral inversion";
3. Frank Knight's emphasis on the significance of the „relatively absolute absolutes", a conception that I personally find indispensable in all discussions on topics such as those that command our attention in this context;
4. Finally, my own oft-repeated emphasis on the necessity of making a distinction in our thinking between the *constitutional* level of discourse, evaluation, or choice and the *postconstitutional* level.[2]

---

2 Only after I had developed an outline of a theme for these remarks did I recall that many aspects of the same argument are contained in an earlier paper that

COMMENT ON THE MORALITY OF CAPITALISM

Peter Koslowski is surely correct when he suggests that we (the set of persons in an existing capitalistic or quasi-capitalistic order) are unwilling to accept fully the individualistic autonomy of the free market (more inclusively, the free society) on moral grounds. We are unwilling to place an explicit positive evaluation on the pattern of results that catallactic interaction generates. This reluctance to attribute what we might call „moral worthiness" to the free economy does not stem either from some failure to value the *liberty* of individual choice that only the capitalistic order allows or from some lack of interest in attainment the highest possible satisfaction of whatever preferences that may exist (*efficiency*). Morally, we reject the autonomy of individual choice that the market embodies because we assign to the market the wholly inappropriate task of generating morally satisfying sets of preferences. In Koslowski's terms, we expect the normative „theory of allocation" to include a normative „theory of preference formation". As Frank Knight was so fond of emphasizing, what we really want is to have „better" preferences.

Knight's suggestion here seems to have empirical support, based both on introspection and external observation. It is, nonetheless, paradoxical in one sense. We are unable to define what set of preferences are „better" . Having long since rejected the „truths" of revealed religion, we cannot construct universally acceptable criteria for „betterness" through the use of scientific-rationalistic procedures. We want what we cannot define, yet, somehow, we know that that which we want is there, behind the curtain that divides present from future. Why, then, do we want „better" preferences?

At this point, Polanyi's notion of moral inversion becomes relevant, the notion that is also expressed in the first of the Koslowski citations at the end of Section One.[3] Polanyi suggests that modern man is characterized by an excess of moral fervor that seeks outlets. With the erosion of emphasis upon personal salvation in a quest for life hereafter, there came in consequence a necessary decline in the internalization of morals. Individual morality was no longer instrumentally required for individually-meaningful purposes. The rationality of science, carried to its limits, suggested that there were no „abso-

---

concentrates on a dispute between Michael Polanyi and Frank Knight. See, JAMES M. BUCHANAN: „Politics and Science", *Ethics*, 77 (July 1967), pp. 303-310. Reprinted in JAMES M. BUCHANAN: *Freedom in Constitutional Contract*, College Station (Texas A & M University Press) 1978, pp. 64-77.

3 Polanyi's conception is discussed in MICHAEL POLANYI and HARRY PROSCH: *Meaning*, Chicago (University of Chicago Press) 1975.

lute" values upon which moral emphasis might be centered. Cannibalism was placed on all fours with Franciscan charity. In this modern environment, persons who seek outlets for their moral passions and who possess no powers of discrimination, become highly vulnerable to exploitation to persuasive leaders who promote the varying „final solutions" of this century. Polanyi holds out the hope that, as the post-Christian residues of moral passion dissipate, modern man may become more willing to accept the autonomous order of the free society.

I find Polanyi's hypothesis to be an interesting one, with perhaps some explanatory value. But I doubt whether we can place much hope in the improvement that it offers us. The Polanyi hypothesis may explain the behavior and the success of some of the zealots of this century (the terror about which Koslowski speaks), and it may, possibly, suggest that the ranks of the zealots may be expected to thin a bit. But it seems to me highly unlikely that man in the late 20th and early 21st century will cease and desist in his moral condemnations of the market order, and that he will at least attain some position of moral neutrality with respect to the freedom of choice that the market embodies. Can we expect our children and grandchildren to accept that pushpin is as good as poetry?

There is, however, a basis in Polanyi's conception that may be turned to constitutive purpose. If we agree that the inherent moral urgings of man will *neither* return to the inward-seeking soul of medieval Europe *nor* fade into moral quiescence, the challenge to the social philosopher stands clear. This challenge is to design institutional structures that will channel these moral urgings so as to satisfy two objectives simultaneously. First, the outlets must be such that those persons, whose whole being exists in their actual or imagined efforts toward furtherance of a „better" world, can experience fulfillment without frustration in some pragmatically limited sense. Second, the very real values of liberty and efficiency that the autonomous order of capitalistic constitutional democracy guarantees must be *preserved.*

To make progress, we must first, at the level of discourse among academic philosophers (broadly defined), such as the participants in this discussion, come to accept and to understand as relevant the basic distinction between the two levels of choice toward which our rhetoric might be directed, the constitutional and the postconstitutional, to which I have earlier, and elsewhere, referred. (A long-continuing personal frustration for me has been the apparent inability of obviously-competent social scientists, who may share my basic values in one sense, literally to understand what I am talking about when I stress the constitutional-postconstitutional distinction.)

What does this distinction, which may admittedly be relevant for political reform, have to do with the morality of capitalism and with prospects for resolving or reducing the moral conflict between the autonomous individualism of the market and the moral urgings of men to live in a „better" society? I suggest that the adoption of the „constitutional attitude" can be of major assistance in any discussion of preference formation. At this point, the Knightian emphasis on the „relatively absolute absolutes" comes to be of critical importance. The normative theory of allocation embodies preferences as they exist, and within this theory these preferences are to be taken as „relatively absolute". Relevant choices within this normative setting are analogous to *postconstitutional* choices; existing preferences are givens in the environment. Within this normative context, pushpin (rock music, Harold Robbins fiction, TV situation comedy) *is* as good as poetry (classical music or Shakespeare).

At a second or „constitutional" level of discourse, however, existent sets of preferences need not be accepted, and, indeed, one of the aims of such discourse becomes effective criticism of such preferences with some view toward „improvement" through appropriate institutional change. Preferences for pushpin are not so good as preferences for poetry, and the social philosopher-cum-scientist has, as one of his central tasks, the design of constitutional-institutional structure that will promote the emergence of „better" preferences (e. g., poetry). His normative effort is grossly, and sometimes tragically, misdirected if he fails to understand the appropriate level of choice for his endeavor, and if, as a consequence, he seeks by resort to the coercive agency of the state to impose his own preferences on others. The moralist who stands outraged at the tastes of his fellows must learn to vent his anger where it belongs, on the institutions that generate the preferences, and notably the educational establishment, rather than on those of the market economy which are, or may be, wholly neutral as between rock and classical music.

Properly conceived and understood, the discussion-debate over the preferred set of constraining institutions within which individuals are to be allowed to exercise their free choices may be characterized by search for a single „truth", waiting to be discovered. Because, however, no one can fully define just what „better" preferences look like, and because the direct objects for adjustment are quasi-permanent institutional parameters, self-interest of the narrow sort will not dominate the behavior of participants in the discussion. Important elements of some Rawls-like veil of ignorance are necessarily present. The relegation of narrow self-interest to a subsidiary role will tend to insure that agreement or consensus is more readily attainable.

My concern here, however, is not whether or not agreement can or cannot be reached on the set of institutions within which the preferences of future persons (including ourselves) will be allowed to emerge. My point is, rather, that it is at this level, and this level only that the exercise of moral passion must be practised. This level of discourse, that which has the normative theory of preference formation as its central focus, can embody severe criticisms of the preferences that are observed.

The moralist who advances such criticisms can, however, and at the same time, accept the latter preferences as they are, as „relatively absolute absolutes" that are to be accorded existential status in the normative theory of allocation. I am not suggesting that the shift in attitudes outlined will be easy to accomplish, or, if once accomplished, easy to maintain. In a game theory analogy, the challenge seems relatively manageable. It seems reasonable to expect that persons might continue to play a game within defined rules while, at the same time, bringing critical judgment to bear on changing these rules. Or, in other words, that they might continue to work to promote desired objectives (continue to seek efficient postconstitutional policy changes) within a defined constitutional structure while simultaneously seeking to promote changes in the structure itself, changes that promise to open up some preferred set of postconstitutional options.

In effect, I am, at least indirectly, interpreting existing preferences as a part of the „rules of the game" in which we are all engaged. If the game is to continue at all, we must learn to accept these rules and play within them, as effectively as we can. But rules are not absolutes; they are subject to criticism, comparative evaluation, and possible change.

In my context, preferences can be shifted, but only through a long-term process of institutional adjustment.

Let us acknowledge the optimistic Enlightenment folly for what it was. If left wholly unconstrained men will not seek only the good, the true, and the beautiful. We cannot so readily escape personal responsibility for the order of society. The free society that is to be worth living in must be *constitutionally constructed.*

Part Two

Evolution and Society
A Critique of Sociobiology

From the Preface to the First German Edition

The book has been developed from a paper given at the conference „Evolution and Liberty“ held jointly by *CIVITAS. Gesellschaft zur Förderung von Wissenschaft und Kunst e.V.*, Munich, and *Liberty Fund Inc.*, Indianapolis, USA, at Munich, Germany, on May 8th to 11th, 1983. The conference united scholars from philosophy, the natural sciences, and economics.

Munich, October 10th, 1983

Preface to the Second German Edition

In revising a book, one used to end up by writing a second, new book. To avoid this risk and since there arose no serious reasons in the five years after the first edition that necessitated a revision of the text, the second edition of the book at hand is published unchanged with only minor corrections of printer's errors.

The reflections of chapter III of the text on the ontology of evolution theory have been further developed in the author's paper „Evolutionstheorie als Soziobiologie und Bioökonomie. Eine Kritik ihres Totalitätsanspruchs“ (Evolution Theory as Sociobiology and Bioeconomics. A Critique of their Claim to Totality), in: R. SPAEMANN, R. LÖW and P. KOSLOWSKI (eds.): *Evolutionismus und Christentum* (Evolutionism and Christianity), Weinheim (Acta Humaniora) 1986. This paper investigates the question whether natural kinds since they cannot form classes - because of the constant change of kinds assumed by the theory of evolution - must be conceived of as individuals. It further develops the idea that the theory of evolution must be considered as a historiography and as a narrative theory. It criticizes finally the monism of evolution theory.

Among the reviews that the first edition of this booklet has found, the review by MICHAEL T. GHISELIN in *Journal of Social and Biological Structures*, 7 (1984), p. 391f., should be mentioned.

Hannover and Herdecke, September 1989

Introduction

The theory of evolution has become a dominating fundamental concept or ontology in the sciences. In the most diverse disciplines, in biological and social theories, „evolution" passes for a basic principle of reality, for a „prima philosophia". It is a sign of modernity to grasp the entirety of reality as an evolutionary process, the principles of which are formulated in the theory of evolution.[1] In this view of the world as a process, a view that understands the universe and all complex systems of life as the result of one evolutionary process, modernity stands opposed to the ancient and medieval cosmology and metaphysics of *ordo* and substance. The theory of evolution replaces a cosmology and metaphysics of the teleological-entelechial order of being and the conceptual primacy of substance or essence. Biology, cosmology, anthropology, even theology[2], are being reformulated in accord with the principle of evolution. Beyond Darwinian biology, the principles of adaption, mutation, and selection are being applied in economics and the social sciences, in theories of socio-economic[3] and socio-cultural evolution[4].

The transition from a substance to a process metaphysics is closely tied to the notion of subjectivity and the history of the subject's freedom in modernity, as is manifested in German idealism by the theories of cultural-social evolution of Franz von Baader, Hegel, Herder, and Schelling. As against the recurrence of the same in substance and the „domination by the overwhelming object" (Hegel), modernity sets forth in its philosophy of history the freely developing evolution of human subjectivity, thereby going beyond Christianity's linear view of history. Yet this passage from substance to subject metaphysics already presented a danger: the danger that an evolution set fully free, no longer recognizing any substantial forms of life, would lose itself in emptiness and arbitrariness or would veer off into the terrorism of a subjectivity liberating itself from all restraints of origin or environment. Hegel's assertion that all depends upon determining subject simultaneously as substance and substance simultaneously as subject, is a response to this danger.

---

1 Cf. KRINGS (1981) and (1984).

2 Cf. the approaches of a process theology following Alfred N. Whitehead.

3 Cf. especially BOULDING (1981).

4 Cf. LUHMANN (1975) and HETTLAGE (1982).

INTRODUCTION

The Darwinian theory of evolution[5] and, ultimately, evolutionary process and system theory succeeded the notion of development in the cultural evolution and subject metaphysics at the turn of the 18th and 19th century. These theories also run the risk of defeating the liberating element in the notion of development when the idea that there exists something like an essence of man becomes obsolete and the unity of the self is replaced by stable patterns of selection and determination.

As is to be shown, this is also true of sociobiology. The person and the person's subjectivity come to be understood as a mere intersection and chance unity of overlapping social and biological-genetic system functions. Subjectivity becomes a mere epiphenomenon of the development of a subsystem that remains stable in time, but has no substantial form, no entelechy; subjectivity and identity of kind are considered the results of crossing-point of interdependent system-factors.

What was already evident in the passage from substance to subject metaphysics, is also the case in this Darwinian and system-theoretical transformation of the notion of evolution and development. In order to retain the moment of freedom in development, the substance concept must be retained in ontology, since, in the theory of evolution, freedom is otherwise conceivable only as chance.

The theory's repercussions for the understanding of the self and for the history of human freedom are ambivalent. Modernity's ambivalence is expressed in the history of the notion of evolution and development. For modernity is as much a history of freedom as it is a process of man's and his subjectivity's becoming obsolete and anachronistic. Man is in danger of being rendered obsolete in the face of the objectivizations of his own nature and self. The sociobiological pretension completely to explain man by means of Darwinian categories as one evolutionary product among others also annuls the idea of human subjectivity and makes the image of man the free and spiritual being appear untenable and anachronistic. But sociobiology jumps to conclusions in making this assertion that are unjustified as will be demonstrated in the following. Phenomena of mind that bar a reductionist explanation will be brought forth against sociobiology's reductionist program.

The equivocacy of the concept „evolution" is evident in the history of the theory of evolution and the contemporary multiplicity of conceptions of evolution. The Lamarckian approach of socio-cultural evolution by learning, the Darwinian approach of natural selection, and the functionalistic, systems ap-

---

5 On Darwin's origin in A. Smith and T. R. Malthus see HIRSHLEIFER (1984).

proach in sociological theories of evolution all have different ontological foundations. While the Darwinian theory of evolution works with the causal-deterministic categories of mutation, adaptation, and selection, the sociological theory of evolution analyzes social systems according to variation, selection, and stabilization. While the basic principle in the Darwinian theory of evolution is genetic selection by way of the survival of the fittest, in the sociological theory of evolution variance appears by means of representation of objects in language and conceptual negation of meaning, and selection by verbal and symbolical communication.[6]

Sociobiology introduces a new moment of monism into this pluralism of evolutionary principles. Sociobiology claims to overcome the pluralism of principles in natural and social evolution and selection and to develop a theory of a unified evolutionary process encompassing both realms of being. The gap separating Darwinian and socio-cultural evolution is to be spanned by comprehensively integrating the social side into a Darwinistically interpreted total evolution. The sociobiological program works out an evolutionary, materialistic monism. This theoretical synthesis based on Darwinian principles, which encompasses the social sciences and economics, will be investigated with attention directed to the question, whether sociobiological monism can justify its own claims and can adequately grasp social and economic phenomena. The investigation shall be based primarily on Wilson's theory of sociobiology[7] and Dawkins's theory of the selfish gene[8].

The inquiry is arranged as follows: After an introduction to the concept and research program of sociobiology in the first chapter, the second depicts the convergence between a biological theory of the social realm and an economic theory of biology, and their mutual integration in a theory of sociobiology and bioeconomics. The third chapter raises considerations on the ontology of sociobiology, critically coming to grips with the view that genes are the ultimate principles of reality, with the epistemological status of optimization theory in sociobiology, and with the extent to which evolution can be reconstructed by optimization theory. The fourth chapter works out the congruities and differences between natural economy and political or social economy and defines the relationship between culturally-ethically and biologically formed purposes and needs. The fifth chapter deals with several unsolved problems of sociobiology; it shows that Wilson's and Lumsden's so-

6 LUHMANN (1975), p. 152.
7 WILSON (1975) and (1978).
8 DAWKINS (1976).

ciobiology cannot meet its claim of creating a comprehensive theory of socio-cultural evolution, because it is unable to develop an adequate theory of intentionality and consciousness. In conclusion, the sixth chapter discusses sociobiology and the theory of evolution as a world-view and theory of totality, and it raises the question as to which criteria of truth we can employ to decide between world-views. Here, as throughout the investigation, it should become clear that a close connection exists between the way we comprehend the basic principles of reality and the way we conceive of ourselves and the social world, it should become clear that ontology and social philosophy cannot be severed.

Chapter I

The Concept and Program of Sociobiology

On the borderline between the social and natural sciences, sociobiology constitutes one of the most recent developments among scientific disciplines. According to the program of its systematic founder and originator of the concept „sociobiology", E. O. Wilson, it is to achieve a synthesis between biology and the social sciences. Just as physics serves as a prior underlying science for chemistry, and chemistry for biology, so, according to Wilson, scientific, evolutionary biology forms the basic science underlying the social sciences and provides them with the explanatory models and fundamental theories of social behavior in the form of genetics and ethology. The simpler, observable behavioral phenomena of the animal kingdom, which are viewed as strategies to maximize survival that have arisen by natural selection, are employed to explain the complex forms of human social behavior and reduce them to the basic functions of maximizing survival.[9]

In addition to this very comprehensive program of human sociobiology, the concept „sociobiology" also designates a more limited empirical research program. A part of biology, it deals with the phenomena, functions, mechanisms, and genesis of social behavior in animals, and attempts to explain

---

9 WILSON (1975) and (1978), LUMSDEN/WILSON (1981).

this behavior with genetic and ecological evolutionary theories.[10] The behavior of animals in their ecological environment, their interaction with their own and other species, with competitors for food, with predators or natural enemies is explained as the result of adaptations to the environment and the genetic selection of the individuals whose strategies and adaptations maximize survival. „Veterinary sociobiology" designates a theory to explain animal behavior, which employs various empirical and theoretical methods: those of game theory, economics, and optimization theory.[11]

Common to these approaches is the neo-Darwinistic, evolutionary point of departure, that „nothing in biology makes sense except in the light of evolution."[12] Evolution is understood as a process in which individuals are selected whose behavior is best adapted to the environment and whose genetic fitness and reproductive success are therefore greatest. In the course of evolution, the genetic information of the „fittest" preserves itself to the greatest extent. According to this neo-Darwinian approach, life forms do not follow the „purpose" of survival of the species or group, but only the „reproductive imperative" of preserving their own genetic information. They tend to maximize their own genetic fitness and to secure the inclusive fitness of their immediate offspring. The priority in securing the survival of relatives correlates with the proximity of kinship.[13]

Sociobiological approaches exclude all types of group selection. The living beings do not act for the good of their species or group, but rather exclusively for the benefit of themselves or their immediate kin (individual and kinship selection). Behavior is explained as the consequence of strategies of self-interested individuals that maximize their utility. By reconstructing the strategies that permit the longest survival of genetic information, an explanation and even the prognosis of behavior in precisely specified environmental contexts become possible. Sociobiology indicates the conditions under which certain strategies of interaction and division of labor among animals will prevail in various environmental constellations. It reconstructs the constitutive conditions of a species and its environment, under which one can expect vari-

10 Cf. FISHER (1930), HAMILTON (1964), TRIVERS (1971), ALEXANDER (1974), GHISELIN (1974), MAYNARD SMITH (1976), DAWKINS (1976) and (1982), BARASH (1977), MARKL (1980a).

11 MAYNARD SMITH (1976) employs models from game theory, GHISELIN (1974) microeconomics, and DAWKINS (1976) optimization theory.

12 Th. DOBZHANSKY, cited by MARKL (1980b), p. 1.

13 Cf. HAMILTON (1964), BECKER (1976), DAWKINS (1976), HIRSHLEIFER (1978a).

ous reproductive forms to arise, such as hermaphroditism and bisexuality[14], or division of labor between castes, such as among the social insects (worker bees, drones, queens).[15]

A theory of behavior which begins with the assumption that all life forms follow the imperative of reproduction and of maximizing inclusive genetic fitness naturally extends to human behavior. In assuming a unified evolutionary process and the validity of genetic-physiological laws for man as well, veterinary sociobiology must be enlarged to human sociobiology. Such a carry-over is propounded in two radically different ways.

The first and most extensive approach to a human sociobiology, that found in the writings of Wilson, Lumsden, and Trivers[16], assumes the same laws apply to human societies as to animal societies. According to this approach, all human societies follow the reproductive imperative and the maximization of inclusive genetic fitness. Social institutions are expedient means that can be reduced to and explained by this purpose (reductionism). Mental and social phenomena are considered to be mere epiphenomena from the genetic-physiological basis of gene-maximization. Society and culture are directly linked to the biological purposes throughout.

The second type of sociobiology, as it is propounded by most empirically-oriented researchers, is more reserved with its theoretical claims. Its proponents admit a distinction between animal and human sociobiology and assume a plurality of phenomenal domains, that is, a difference between genetic-physiologically conditioned phenomena and phenomena determined by the conceptual and linguistic capabilities of the human mind. Thus, on the problem of a carry-over from veterinary to human sociobiology, Markl writes: „I remain unconvinced that cultural evolution is merely a continuation of natural evolution with other means, and that cultures evolve primarily as mechanisms for maximizing Darwinian fitness of their members or even of their genes. There are more ‚optimization criteria‘ for cultural evolution than reproductive success."[17] Ghiselin assumes that the reproductive imperative, the laws of genetic competition, and the laws of competition over reproductive resources apply in human sociobiology as in the animal kingdom;

---

14 Cf. GHISELIN (1974).

15 Cf. OSTER/WILSON (1978).

16 According to TRIVERS (quoted in *Time,* 110 (1976), p. 54), sociobiology will one day take over the disciplines of „political science, law, economics, psychiatry, and anthropology".

17 MARKL (1980a), p. 8. Cf. MARKL (1976), pp. 25 and 36, and BUCKLEY (1977).

however, the human mind introduces a new quality to human society, because it weakens reproductive competition. It diverts individual energies to goods that are not dominated by rivalry over consumption. With these goods, the consumption of resources by one person does not reduce the reserve of resources that remain for another. The conflict resolving and competition reducing qualities of art, religion, and play alleviate the intensity of reproductive competition over scarce resources, with the consequence that animal and human sociobiology cannot be identical.[18]

It is obvious that „sociobiology" is no unified methodological approach, but rather a program with differing ontological and epistemological starting points. This program can be reductionistically and monistically propounded as a materialistic theory of the evolution of the entirety of reality, together with the accompanying global claims. But it can also be grounded as a pragmatic, hypothetical research program investigating the genetics and evolutionary biology of both animal and human societies, whereby the extent to which it will prove fruitful in the human domain is still unknown. The claim of founding an adequate ontology of social phenomena with theories of evolutionary genetics and ecology is then renounced or at least postponed until empirical research shows further progress.

Discriminating between a monistic sociobiology propounding a „worldview" or „Weltanschauung" on the one hand and sociobiology as a hypothesis and strategy for research into a comprehensive account of behavior on the other is not always easy for two reasons. For one, only a few authors, like Wilson and Lumsden, explicitly declare their ontological principles and their claims of total explanation, or explicitly propound the monistic ontology of

18 GHISELIN (1974), p. 262f.: „Food may run out, there may not be enough mates to satisfy the demand, but our stocks of laughs can never be exhausted. So it is with the other higher manifestations of our inner lives, including our aesthetic, intellectual, and purely religious activities. We do not compete for their objects, and nothing is lost by giving them away... We have evolved a nervous system that acts in the interest of our gonads, and are attuned to the demands of reproductive competition ... Nonetheless, an ability to come to grips with the world as it really is can hardly be dismissed as maladaptive. Above all else, the truth has ethical significance ... If, as we have some reason to think, selection within the context of society has somewhat elevated our baser sentiments, and if, as the facts suggest, self-interest and common welfare are not fundamentally beyond reconciliation, we can reasonably hope to develop ethical standards consistent with biological reality."

a unified process of evolution. They restrict themselves to partial models sociobiologically explaining particular human behavioral patterns, such as sexuality or territoriality. Secondly, some authors propose the thesis that animal and human social behavior are identical and a unified sociobiological explanation is possible. They employ metaphors from the human social realm to sociobiologically describe and explain animal behavior. Thus, for example, Dawkins largely limits himself to suggestions and allusions in transferring experimental results and theorems of sociobiology to man, but never explicitly maintains that human and animal sociobiology follow the same laws. Since, however, the sociobiological description and explanation of animal behavior always proceeds with anthropomorphic categories and metaphors drawn from human social behavior, the transfer of what is said about animal behavior to human society is suggested to the reader.

The failure to openly express and ground the analogies between veterinary and human sociobiology is both ontologically and epistemologically unsatisfying. Yet this also has sociological and ideological side-effects, because social theories always exercise a legitimizing function in society and are a type of societal self-interpretation and self-design. Social theories are never mere descriptions, but are always simultaneously interpretations and definitions of the social world, which, not all too seldom, become self-fulfilling prophecies.[19] This is also true of human sociobiology and the transfer of research results in evolutionary genetics and evolutionary ecology to man. When human society interprets itself according to the sociobiological model and takes the reproductive imperative to be the ultimate purpose of its social teleology, this cannot remain without consequence for the consciousness of the society's members. The introduction of social theories in a social system differs from the introduction of physicalistic theories concerning extra-social reality into a society, because the latter bears fewer effects on the society's self-definition. Therefore, an inquiry into the scope of sociobiology must also consider these practical, social effects on human self-understanding.

The adoption of sociobiological, evolutionary theories in our culture is not merely a scientific question as to the correctness of hypotheses, but also a normative question of social philosophy, whether we want to and ought to understand ourselves by the sociobiological model. Whether sociobiology can found an adequate self-understanding of human societies is neither a purely normative or moral question, nor a purely empirically falsifiable question. The way a society defines itself simultaneously constitutes an aspect of this

19 Cf. MERTON (1957).

society's reality. As W. I. Thomas the sociologist of symbol interactionism, demonstrated, definitions of social reality are real in their consequences.[20] Reductionism perceives social reality as it has previously defined social reality, and, upon general acceptance of the reductionist view, the society comes to resemble the model the reductionist has methodologically preconceived: this is the problem with all sociological reductionism. The following inquiry into sociobiology will therefore be double: first, an inquiry into the ontological basis of its research approach and, secondly, whether its description of social reality and economy can be adequate to our self-understanding.

That sociobiology is not an empirical theory of society, but a wider theory implying social-philosophical consequences becomes most evident when it claims to be constitutive of a world-view and to create a theory of totality comprehending the whole of reality: in E. O. Wilson's *On Human Nature*. With its claim to totality and global character, Wilson's conception of sociobiology takes up genuine philosophical questions. Wilson's monism and „scientific materialism" is developed in explicit antagonism and rivalry to religious and philosophical world views. Because Wilson, in contrast to other authors, presents his model of sociobiology as a theory of totality, not merely as a research hypothesis of moderate scope, it cannot be ignored by philosophy. The philosophical provocation is apparent where Wilson's assertions go beyond empirically secured research results and where his conclusions are not shared by all sociobiologists. Precisely the scientific theories that constitute a world-view, functioning as cosmological and social orientations, develop the greatest social and ideological force.[21] If, after the end of metaphysics as declared in analytical philosophy and the philosophy of science, philosophy does not wish to lose its task of constituting and criticizing world-views, or to surrender the task to the theory of evolution, then it must face the challenge of sociobiology. Therefore, after discussing impact of sociobiology on social philosophy and economics, its metaphysical and religious dimension is also to be investigated.

---

20 THOMAS (1928), p. 81.

21 Cf. MARTEN (1982), PÖRKSEN (1983).

Chapter II

Sociobiology as Bioeconomics

1. The Economy of Nature

Evolutionary ecology and sociobiology view organic nature as a continually changing web of adaptation and interaction, in which the life forms both compete for scarce resources serving life and reproduction and must make the most efficient use possible of these resources. The ecology is understood as an economy of nature. The same laws apply in this bioeconomy as in human economy: „One universe, one economy, one economics“.[22] Therefore, theorems and laws of economics, especially microeconomics, can and must be employed in explaining the evolution and ecology of plant and animal behavior.

Michael Ghiselin has presented the most comprehensive integration of economic theories in ecological and sociobiological questions to date with his work *The Economy of Nature and the Evolution of Sex.* For him, the economy of nature is completely a competitive market economy. Natural and sexual selection is a manifestation of competition for scarce resources, such as fertile females. With this competition, only those who make efficient use of their own resources and capital and of environmental resources can survive and reproduce. Species and individuals, predators and prey compete for food and energy for their reproduction. The ecology is an economy in which the laws of the market are valid. The extent of the market determines the degree of the division of labor and specialization, the population density in a territory fixes the extent of competition and the degree of social differentiation in castes. The law of diminishing returns of a factor of production determines the optimal size of a firm or an enterprise, e.g., the size of a nest and the number of offspring raised in it. Diminishing returns from one activity and increases in productivity by specialization effect the differentiation of the sexes and the rise of different forms of reproduction.

Ghiselin's work, which is impressive not only for its profusion of empirical material, but also its eminently speculative aspect, aims to explain the evolution of different forms of sexual reproduction. He convincingly shows

22 GHISELIN (1974), p. 212. Cf. RAPPORT/TURNER (1977) for an overview of the use of economic laws in ecology.

that the economy of nature is not an economy of linear optimization or simple maximization of gene replicates, not an economy of thriftiness, but an economy of innovation and prodigal change. Sexual reproduction, that is, the recombination of hereditary information in DNA from the hereditary material of two individuals is, to begin with, wasteful. It costs energy, material, and time for seeking a mate, and interferes with feeding and growth. However, it is a means for mobilizing genetic variability and the ability to adapt to changing environments. The continually varying combination of DNA in sexual reproduction enables adaptation to the environment and exploitation of new resources. According to Ghiselin[23], the flow of genes in recombined DNA can be compared to the flow of capital and technological knowledge, a flow which continually channels capital and know-how into new productive uses, and renders possible innovative reshuffling of existing capital stocks.

Ghiselin is nevertheless far from propounding a metaphysical economism which sees in nature a single maximization process, and the best or most efficient of all possible worlds. The natural economies do not necessarily attain maximum efficiency. They do not maximize particular targeted quantities; rather, particular ecological interactions can be explained with the aid of economic regularities. „The idea that natural economies necessarily evolve in the direction of efficiency is every bit as much a metaphysical delusion as the notion that good will inevitably result if only government lets capitalists do as they will."[24] Economic laws are means to understand and partially explain the interaction and behavior of living beings, not, however, a metaphysical magic key to the inner secrets of nature that would permit us to understand all interactions of life forms as moments of a comprehensive optimization process.

2. Monocausality and Theory Imperialism

Sociobiology is transformed into a metaphysical biologism and economics likewise into an economism when their paradigms are understood as conclusive theories of totality and the structure of their analytic apparatus considered to be the fundamental structure of reality, the ultimate principles of being, in biology the maximization of genetic fitness, and in economics utility maximization. Both uses render *one* explanatory principle absolute, mak-

23 GHISELIN (1974), p. 54.
24 GHISELIN (1975), p. 88.

ing it into a theoretical monism and a monocausal explanation of reality. Such ontologizing of a scientific analytic instrument and explanatory principle into the fundamental principle of being is scientistic, because it fails to do justice to the wealth of phenomena of natural and social life. Theoretical monocausality is an imperialistic attempt to make the perspective of one science the measure of all other disciplines.

The theory imperialism[25] in contemporary biology coincides with that of economic approaches. In the last decades, economics has likewise shown a tendency to expand its paradigms, attempting to explain all social domains by means of the theorems of utility maximization and economic interaction (economics of the family, of marriage, economic theories of politics). Although economics can show some significant successes in explaining human behavior, it has not succeeded in producing a unified, universal explanation and prognosis of human behavior, because the concept of utility in the fundamental principle of microeconomics, utility maximization, is too open and formal. In all areas where the principle of utility maximization is applied, it must first be clarified what „utility“ is to mean and which utility is to be maximized. The goods from which utility is to be drawn, the arguments x_1 ... x_n of the utility function $U = U(x_1 \ldots x_n)$, are neither given nor are they anthropological constants. Not all human actions can be uniformly understood as the maximization of certain variables, because needs and definitions of utility display too large a social and cultural variation.

Utility maximization cannot be transformed into a technical problem of optimizing fixed variables. In this situation, which proves unsatisfying for a unified, monocausal economic explanation of action, sociobiology must seem a welcome ally of economics, because the sociobiological approach permits the translation of the subjectivism of economic utility maximization into a bioeconomic, sociobiological objectivism of maximizing genetic fitness.[26] A unified theory on the behavior of life forms, a „unified field theory“ of behavior, so to speak, appears possible, if there is a common denominator for all optimization strategies of life forms and a „final purpose“ of all behavior and action. The hypothesis of the maximization of genetic fitness renders it possible to interpret all other purposes of living beings from the amoeba to man as expedient means to this final purpose, as means promoting genetic

25 BOULDING (1969), p. 118.

26 As in BECKER (1976). HIRSHLEIFER (1978a) and (1984) makes a stronger distinction between „natural economy“ and „political economy“. Cf. also the overview by LEHMANN (1982).

fitness. In this purpose, sociobiology and bioeconomics unite to form a single science of the optimization strategies that all life forms employ to secure the survival of their genes. Knight[27] criticizes utilitarianism for transforming ethics into economics with its criterium „utility", and transforming the ethical problem of selecting between purposes and values into an economic problem of selecting means to the given purpose „utility". This reduction of the problem of decision between purposes to the economizing of means to *one* purpose applies all the more so to the bioeconomics of the selfish gene. In this view, the ethical problem of the proper life is transformed into the problem of optimally engaging all means and all manifestations of life for the sake of gene survival.

3. The Gene as Investor

The most radical form of this sociobiological monism of optimizing gene survival is presented by Dawkins. Living beings, according to Dawkins, are survival machines of selfish genes. „We are survival machines -- robot vehicles blindly programmed to preserve the selfish molecules known as genes."[28] Genes furnish living beings with an optimization program for mastering complex situations in the outer world and attend to securing an inner and outer environment that is optimal for the final purpose, the survival of genes. This program regulates the self-replication of these genes over generations. The arguments of the economic utility maximization function $U = U(x_1, \dots x_n)$ can now be clearly determined. The rule of utility maximization Max! U can be translated into the rule: „Maximize your genetic fitness including that of your offspring", Max! W_p^*. The economic utility function with a number of unknown variables $U = U(x_1, \dots x_n)$ can be replaced by a utility function with „only" one unknown variable $U = U(W_p^*)$. The inclusive genetic fitness of the parent W_p^* depends on the parent's genetic fitness including the genetic fitness of his offspring weighted by the degree of kinship[29], and it is described by the inclusive fitness function $W_p^* = W_p + r_{po}W_o$. r_{po} is the coefficient of the degree of relationship between parents and offspring, and W_o the genetic fitness of the offspring. The consideration of relatives' genetic fitness in an individual's decisions increases with their genetic

27 KNIGHT (1935), p. 34.
28 DAWKINS (1976), p. X.
29 Cf. HIRSHLEIFER (1978a), p. 241.

proximity to the parent, with the magnitude of the kinship coefficient r_{po}. (For a parent's own children r_{po} = 0,5 or 50 %.) In optimizing the life plan from a bioeconomic and genetic viewpoint, the task of utility maximization Max! U (x_1, ... x_n) is converted into the much simpler task Max! U (W^*).

Such a model is tempting in its simplicity (monocausality), scope, and elegance. It is a theory of outright cosmic dimensions, because it unites natural and human history under one single purpose, the purpose of the survival of genes. The lifelong task and sole purpose of man is maximizing the survival of his genes, the purpose of mankind is preserving the gene pool. Human history is no longer viewed as the continuation of natural history, but rather as being identical with the history and purpose of nature. Natural history and human history became indistinguishable, because both concern exclusively the preservation of genetic information. Against the categorical imperative of gene survival, all other imperatives become hypothetical imperatives.

Consistently set forth, this would lead up to the justification of cannibalism, as Jünger feared: „First steps, transitions to intelligent cannibalism suggest themselves, often even unconcealed. Every purely economic view will necessarily lead to it".[30] According to sociobiology, the optimal use of food resources requires that animal mothers devour the portion of their offspring that does not serve the survival of their genes because it hinders the survival of siblings.[31] Authors like Dawkins and Wilson shrink from a transfer of this sort from bio-economy to human beings.[32] Nevertheless, it would have to be concluded in sociobiology, if cannibalism in families promotes the survival of genes. Wilson holds to human rights, but his justification for this goes beyond his own sociobiological theory.[33]

Dawkins' approach grants genes something on the order of intentionality or consciousness. He falls into a genetic animism, attributing perception and devision to genes, largely surpassing the capacity and speed of Darwinian mechanisms for selection. Such an animism breaks through when Dawkins employs economic analogies, and leads to an ontologizing of economics and the concept of rational pursuit of goals. Genes are viewed as „portfolio in-

---

30 JÜNGER (1958), p. 160.

31 DAWKINS (1976), pp. 144ff.; GHISELIN (1974), p. 231.

32 DAWKINS (1976) is especially inconsistent on this point: on the one hand we are blind robots used by selfish genes (p. X), on the other hand, we can upset the plans of genes (p. 3), and cannot derive ethics from evolution!

33 WILSON (1978), p. 187.

vestors on the stockmarket", whose stocks or enterprises are the survival machines in which they invest.[34] Genes act as profit maximizers in programing the construction of their survival machines; they optimize profit in their strategies like entrepreneurs; they measure their „investment policy" against the „gold standard of evolution, gene survival"[35]. Accordingly, for Dawkins, genes possess a learning ability corresponding to that of profit-maximizing individuals. The difference in the dimension of time for genetic and for mental learning is obscured and annulled by Dawkins in an animistic fashion.[36]

This is the case, for example, in his explanation of menopause as the result of selection to the benefit of grandchildren. According to Dawkins, it is more advantageous for the genes of an elderly woman to invest in grandchildren rather than in her own children. Thus, the gene for „altruism toward grandchildren" has evolutionarily established itself in the gene pool.[37] A woman cannot invest the maximum in her grandchildren if she continues to have children of her own. This is not an evolutionary-genetic explanation, but an economic-rational explanation for reproductive behavior. The life expectancy for women rose from 35 to about 70 years only in the last 100 years, and it can hardly be expected that menopause at the age of 49 established itself by the way of genetic selection within these 100 years. This example shows that Dawkins' anthropomorphic application of economic models to genetic evolution inacceptably oversteps the Darwinian model of selection. Subjective and rational qualities are attributed to genes; genetic learning

34 DAWKINS (1976), p. 60.

35 DAWKINS (1976), p. 133.

36 DAWKINS (1976), p. 132, p. 50 and else where, excuses his use of anthropomorphic categories as being an abbreviated manner of speech, permitting a simpler representation of complex connections: „... it is often tedious and unnecessary to keep dragging genes in when we discuss the behavior of survival machines. In practice it is usually convenient, as an approximation, to regard the individual body as an agent ‚trying' to increase the numbers of all its genes in future generations." It can be objected, that the metaphors are not insignificant. Rather, they conceal the fact that the anthropomorphic capabilities and subject qualities of genes cannot be reconstructed with theorems from the theory of selection alone. The metaphors allow the use of teleological concepts while maintaining the appearance that the model is a thoroughly mechanical one. For a critique of the use of metaphors and quotation marks in modern biology cf. LÖW (1980), p. 283.

37 DAWKINS (1976), S. 135.

is not distinguished from social learning. The economizing gene comes to be the actor behind the acts.

Chapter III

On the Ontology of the Theory of Evolution

Dawkins' approach raises the ontological question: what are we to consider the ultimate principles of being, the form and its realization, or the preservation of genetic information. By shifting the phenomenon of intentionality to genes, Dawkins is maintaining that the preservation of genetic information is the final cause of all existence. The preservation of genes stands higher than the purpose of preservation of the individual or even the species. The form of the organism is not maintained as a total program; rather, independently from the organism as a whole, the elements of the genetic code aim to increase their fitness for survival. „The slightly selfish big bits of chromosome" strive to preserve themselves.[38] Not only is the concept of species nominalistically broken up, as in Darwinism and Neodarwinism[39], but even the concept of the individual falls into a super-nominalism. The identity of him who secures his genetic fitness can no longer be fixed either in language or empirically. It makes no sense to speak either about the history of the species humankind or about that of an individual human. One can speak about the history of the gene pool, which uses machines that demonstrate a remarkably high fidelity to form and that give these same machines an artificial word as a name, i.e., man. The concepts of species and individual possess no reference, they do not designate anything real. „Man" is merely the summary for a symbiosis of machines that serve the preservation of information of small bits of chromosomes.

38 DAWKINS (1976), p. 35.
39 Cf. MAYR (1963), p. 337. For criticism cf. FELLMANN (1977), p. 286.

EVOLUTION AND SOCIETY. A CRITIQUE OF SOCIOBIOLOGY

1. Gene Survival and Program Preservation as the Entelechy of Nature?

The supernominalism reducing the individual to an aggregate of survival machines is led *ad absurdum* by Dawkins: perhaps we are not individuals at all, but multiple organisms whose genes make use of our awareness of identity in order to secure their preservation in a symbiosis.[40] Genetic supernominalism, according to which small units of genetic information, not species or individuals, are the ultimate ontological unity of reality and life, dialectically turns into an abstract pseudo-Platonic essentialism or super-idealism.[41] According to Dawkins, the being and survival of small units of information which seek out and „exploit" their bodily carriers constitutes the essence of life. The corporeal, formed being of individuals comes to be an epiphenomenon of the actual being of genes. Such a genetic „idealism" raises the question: why does something take on shape and being at all, when its teleological purpose is the survival of something entirely different, invisible, and without shape. If survival of genes is the purpose and this survival program steers the reality of life, then the world perceived by us is to a great extent non functional or luxurious, because it has form. We human beings aim at the realization of forms, not at abstract idealisms. It would be more economical for genes to eternally swim in a primordial soup and preserve their informational content in the state of potentiality, without ever converting this information into formed reality. The *realization* of the information of DNA in the form of the individual is ontologically superfluous, because the purpose is alone the preservation of this information. If only the blueprints are to remain, it is superfluous and inexpedient to build the cathedral. Considering the genes to be the real agents and locating the *causa finalis* and

40 DAWKINS (1976), p. 196.

41 Cf. GHISELIN (1974), p. 38: „Treating a gene or a population as if it were an Aristotelian ,that for the sake of which a thing exists' *sweeps* the whole problem (of teleology) under the rug". MASTERS (1979), p. 251: „For contemporary biology as for Plato, the visible animals can be said to 'represent' the information or *eidos* which proceeds them in time and space". This is not true in the case of Dawkins, because the *eidos* is divided into small bits of chromosomes. Masters' comparison of the genotype with the Aristotelian concept of the soul as *forma corporis* is also inaccurate, because precisely the constancy of this form is abandoned in sociobiology.

causa efficiens of all existence in the individual gene confuses book-keeping with causality.[42]

Dawkins' reductionist definition of life as the replication of genetic information makes his theory implausible for optimization theory. Life and information could be preserved with less expense, because the replication of information could be accomplished more expediently and economically without organisms. Why does man - or why do the genes by way of him - generate a human; why instead don't small bits of chromosomes generate other small bits? Since Aristotle, generation has been conceived as endowed with form and connected with species; the concept of generation is difficult to apply to genes. Thus, our experience also distinguishes it from ways of making and of copying or replicating. Dawkins should be translated thus: A human begets human, in order that a gene replicate itself. The gene replicates itself by causing a human to beget another human, who possesses half of the same genes. This is not an economical procedure. It is like a copy shop which, for producing copies, always copies the copying machine as well, obtaining copies only half corresponding to the original.[43]

As idea and pure information, genes could be eternally *in potentia* without the body and without exposing themselves to perishing by taking on shape *in actu*. If, on the other hand, the gene is conceived as an entelechial information, which must become real, then the purpose of the living being is the form and its realization, not the survival of potential information. The inversion of the entelechy from the form to the survival of this form's genetic information is ontologically implausible and logically contradictory. Something eternal and potential, the gene, is to realize itself in something finite and real, the form. But the finite thing, as finite and realized, is not the purpose; rather, the survival of the potential as potential remains the final pur-

42 GOULD (1984), p. 6: „I strongly disagree with Dawkins, since I feel that he has confused bookkeeping (which may be done efficiently in term of genes) with causality." (English original text of the German translation)

43 DAWKINS (1976), p. 46, himself raises the objection that sexual reproduction calls into question his thesis that an individual attempts to maximize the number of surviving genes. In the recombination of DNA only half of the genetic code is preserved. However, according to Dawkins, this demonstrates that the concern is not preservation of the *individual's* genetic code in entirety, but rather the survival of bits of chromosomes, for which the preservation of the DNA chain as a whole is irrelevant. The question remains open, why crossing over takes place at all, if only the preservation of little bits of chromosomes is the goal.

pose. The most extreme reversal from the teleology of form to teleology of information, the reversal from completion of the potential in the real *(energeia)* to preservation of the potential as potential *(dynamis)* is found in the economizing and selfish gene. The potential and realizable strives to preserve itself as a potential, not real.

The question arises as to what genetic information can actually signify, if its realization in form or its conscious representation for us, its being-for-another and its entelechy in the completed form are not considered to be its actual „for the sake of which". As Spaemann and Löw have demonstrated[44], information and ideative content are always for a consciousness and do not constitute in themselves mental content. In Hegelian terms, the genes have a „being in itself" but no „being for itself". They are, in themselves, indifferent to their realization or non-realization. Only to us does their realization or non-realization make a difference. Only realized information is conceivable and expressible as information. The genes cannot aim at their realization, because their reality is potentiality. They are not the entelechy of existence; rather, their taking shape in the realized form constitutes the entelechial structure of reality.

After what Freud calls the three affronts to humankind's naïve self-love: the Copernican revolution, the Darwinian theory of descent, and the psychoanalytic degradation of the ego, Dawkins' would be the fourth and last. It would not only expel humankind from the center of the cosmos, deprive it of its uniqueness among the species and of its consciousness of the ego, but would furthermore reveal its bodily individuality and its self-preservation to be false illusion.

Dawkins's theory is the final form of ontological nihilism - nihilism, because the being of a human is no longer worth preserving as itself. It merely serves the preservation of something which itself is not real but only possible. The end of being is the preservation of a program. The survival of the program, no longer the life itself as realization of this program, is the purpose.

---

44 SPAEMANN/LÖW (1981).

2. Critique of Optimization Theoretical Reconstructions of Evolution in Sociobiology

The model of the selfish gene unifies biology and economy by functionally relating all actions of life forms to the maximization of inclusive genetic fitness. With the aid of maximization models, population-biological interactions and social structures are reconstructed. Reconstruction in accord with the optimization or maximization model facilitates the investigation of complex interrelations. It can take the place of empirical research. Because it considerably reduces the effort for inductive observations and experimental simulation, it reduces the expenses of empirical research. Yet it remains an „as-if explanation" and a transfer of human rationality to an ontologically distinct domain. We attempt to understand the social insects *as if* they rationally maximized their inclusive genetic fitness.

In the first place, the still unanswered question arises, what ontological status the maximization principle has, whether it is a physical-energetic principle, or a mental-rational principle that cannot be carried over to instinctive behavior. Samuelson propounds the view that the maximization principle must be conceived purely mechanistically in biology. He criticizes the maximization models in biology for being too finalistically formulated and using teleological concepts. Maximization problems should not be formulated with teleological metaphors of intention, as if nature maximized something. Rather, maximizations should be understood as purely physical reciprocal effects between systems, whereby patterns arise that can be understood *ex post facto* as maximization problems. According to Samuelson, this system analysis does not require a „preexistent" problem that „needed" a solution.[45] As an economist, Samuelson criticizes biology for persisting in arguing too teleologically in maximization analysis. According to him, the maximization-minimization principle is an application of Newtonian physics. Motions in nature obey Maupertuis' principle of least action. According to Samuelson, maximization problems in sociobiology, such as the interactions between populations in one territory, must also be mechanistically explained by the Maupertuis-Euler-Lagrange-Jacobi principle of least action.[46]

---

45 SAMUELSON (1978), p. 183.

46 *Ibid.*, p. 182: „If biological behavior equations happen to provide the maximizing solutions to some definable problem, that does not mean that any mind must know in advance anything about that problem (including anything about its very existence or non-existence). Apples fall in Lincoln-

Nevertheless it is doubtful whether the explanatory achievements of sociobiology can be attained with purely mechanistic models, because models that work with anthropomorphic categories usually are simpler and contain elements of rational foresight and anticipation, without which the maximization principle would lose its explanatory and predictive power. In contrast to the question, what optimal path must be taken by an apple that wants to reach the earth from a tree, the question what reproductive strategy selfish genes in particular populations and environments should „choose" can hardly be explained with the principle of mechanistic optimization as interaction between energetic systems without the equations becoming much too complicated. It is understandable that sociobiological theory is less mechanistic than Samuelson the economist calls for.

The teleological element in the theory of maximization can already be shown at its historical beginnings, in Maupertuis's least action principle. Maupertuis's principle of least action is founded in his metaphysics and understood as a proof of the existence of God. Nature obeys the principle of least action, since God keeps all things in motion in such a way that motion or action defined as *msv*, as mass multiplied by distance and velocity, is minimized: „No force, no quantity which could be regarded as a cause in the distribution of motion subsists inalterably but there is one which, produced anew and created so to say at each instant, is always created with the greatest possible economy ... This force is what we have called the action."[47] All motion stops as soon as God ceases to evoke it in accordance with the least-action-principle. This principle represents, according to Maupertuis, „the universe's dependence on and need for the presence of its creator; it likewise shows that this creator is as sage as he is powerful."[48] It shows that the motions in nature are neither eternal nor independent but subject to a power which produces them and augments them, diminishes them and destroys them *in the most economical way*.

Maupertuis radicalized Newton's assumption that God intervenes in the mechanical machine of the world from time to time[49] by adopting Male-

---

shire to minimize Hamilton's Action integral, and to minimize the Maupertuis-Euler-Lagrange-Jacobi-Least-Action expression. They did so before Newton. Apples don't know they are doing so."

47 MAUPERTUIS (1759), p. XXVI.

48 *Ibid.*

49 See LEIBNIZ, *Letter to Clarke*, quoted in GUEROULT (1967), p. 233: „Mister Newton and his followers still have a funny opinion of God's work. According to them, God is forced to set his watch from time to time. Otherwise, it

branche's theory of occasionalism. *All* motions are sustained by God economically. The impulse is the occasion on which God intervenes in the world: The impulse is only the occasion which determines God to move the body. Since Maupertuis's least-action-principle is consistent only under the assumption that it is God who keeps nature on the path of least action, optimization or least action principles cannot serve as the basis for a maximization theoretical reconstruction of evolution in a mechanistic model of biology. Maupertuis himself elaborated a finalistic, monadic theory of biology and evolution which is Leibnizian and vitalistic rather than mechanistic.

Maupertuis's model of mechanism is ultra-teleological since it assumes that God causes the motions in nature to follow the least action-principle. Maupertuis, therefore, demonstrates that, in the history of ideas, the mechanization of nature in a non-evolutionary model presupposes the deteleologization of the inner nature and, at the same time, causes an externalization of the *telos* to a subject with rational foresight: God makes the mechanism of nature run with least action, but without a built-in teleology, by intervening at each impulse. Without the assumption of the external maximizer it is, however, unclear why minimum/maximum principles should hold true in the totality of nature and it is furthermore untenable that local maxima or even optima form a composition to overall optima. In a purely mechanistic model of nature, there are only local optima.

In conclusion from this comparison of Maupertuis and Samuelson it is not feasible to extend and universalize the maximization principle to an ontological principle without accepting the metaphysical presuppositions of Maupertuis's theory and the least-action-principle. Maupertuis' *metaphysical* grounding of the least-action-principle indicates the profound analytical problem of maximization theory if it is applied to total reality and this means if it is applied as a metaphysical principle: how is a total structure whose interactions follow optimization principles conceivable and sustainable without an external maximizer and energy supplier? The universalization of the maximization principle from a maximum defined for a local context to the context of totality is the inductive conclusion and generalization from a well-defined to a total context, a conclusion which is neither empirically nor experimentally falsifiable or provable.

---

would cease to move. God has not enough foresight to make an eternal motion." (my translation)

The question as to the ontological status of the economization and optimization principle cannot be conclusively cleared up here.[50] Tied to the problem of the ontological status of the optimization principle in organic nature is an equally significant epistemological objection to optimization models. Reconstructions or as-if explanations of optimization are *ad hoc* explanations. They resemble an *ad hoc* explanation insofar as the explanatory context is defined more or less „*ad hoc*" and arbitrarily. The complexity of a situation which an economic gene faces must, in the reconstruction of optimization, necessarily be so greatly simplified that the description of the situation constituting the antecedent conditions for the explanation possesses only a distant resemblance to the reality. „A simple theory about phenomena which by their nature are complex is probably necessarily false, in any case without a specified *ceteris paribus* clause, and after the latter's complete formulation the theory would no longer be simple".[51] These same points of critisism apply to the reconstruction of genetic selection as a process that optimizes gene survival just as they apply to economic maximization models. These models, if they are to reproduce the reality, require a knowledge that is not available.

For an optimization model to correctly describe the reality, the following conditions must be satisfied:

1. the original situation must be correctly described,
2. the set (number and kind) of possible strategies must be known,
3. the optimization criterion, i.e., the variable that is to be maximized, must be well defined,
4. the constraints must be exactly described.[52]

These conditions are satisfied only for „well structured", clearly delimitable decision problems, such as technical optimization in engineering. For complex economic contexts like strategic management decisions, the maximization model can give no unambiguous solution or directive for action due to uncertainty about others' strategies and due to the complexity of environmental conditions and side-effects of strategies in the future.[53] Optimization models have therefore been replaced in economics by satisficing models.[54] Satisficing models are more realistic and more robust. They require attain-

50 Cf. KOSLOWSKI (1985) for a further discussion.

51 HAYEK (1972), p. 16.

52 On the critique of optimization models cf. OSTER/WILSON (1978), p. 297, and LEWONTIN (1979), pp. 8f.

53 Cf. ALCHIAN (1950) and KOSLOWSKI (1982), pp. 230ff.

54 SIMON (1978).

ment of a predetermined satisfactory, not the optimal, level of the target variable.

The decision-theoretical objection with respect to the definition of the original situation and to the effects of action is even more significant when the reconstruction of biological populations according to rules of gene optimization is concerned. Rational foresight is not given in selection models, the gene's decision situation is delimited from without by the observer, and constraints and strategies are not given or fixed, but are also reconstructed by the observer. These economic objections are joined by objections in a narrower sense biological. Optimization arguments in biology assume that, if a particular phenotype is optimally adapted, the fitting and corresponding genotype arises, which has not been proven. Some optimal solutions are mixtures of phenotypes that are dynamically possible only under extremely restrictive genetic conditions. Lewontin convincingly summarizes the problem: the dynamics of natural selection does not include rational foresight, and there is no theoretical principle guaranteeing optimization to be the consequence of selection.[55]

In biology, optimization arguments ought to be employed with the utmost caution if they are not to be an inadmissable carry-over of anthropomorphic arguments to the theory of selection. Selection and optimization arguments cannot be simultaneously valid. Either the selection model holds -- then optimality by complete foresight cannot be assumed -- or the optimization model holds -- then the gene must be conceived as intentional and foreseeing and nature as teleologically leading to optima. In a consistent optimization model, selection would assure only that a mechanistic interaction between the optimization strategies of the species arises but not an equilibrium in the manner of a general market equilibrium between self-interested strategies of economic subjects with rational foresight as it is developed in economic theory.

Optimization models are applicable only to well-defined problems. If the original position, the constraints, the possible strategies, and the objective function can be well defined, an optimal strategy can be identified. It is nevertheless not possible to specify optimal solutions for the universal context of nature. As a total theory of transspecific evolution, therefore, the theory of the selfish gene, which lays the foundation of sociobiology, is not applicable. Whether the reality of nature is the result of optimizing behavior by genes is a question that transcends experience and is immunized against every

---

55 LEWONTIN (1979), p. 12.

possible empirical and introspective falsification. Even the optimization criterium is unclear. Is the stock of chromosome parts optimized, as in Dawkins, or of entire DNA chains?[56] Is the present stock or an intertemporal stock of genes optimized? Due to our lack of knowledge and the exponentially increasing complexity of larger systems, we can specify optima only for regional contexts, we can only indicate local optima.

The origination of an entire species cannot be comprehended as an optimization process. It remains indeterminable in optimization theory. All that can be said is: if a population or an individual seeks to secure the survival of its genes under given environmental conditions and given constraints, it must behave in such and such a way. How it comes about that it finds itself phenotypically and genotypically in such a decision situation cannot be described or theoretically reconstructed as a result of previous optimizations. In optimization theory the description or reconstruction of the evolution of a species must either be incomplete, because we lack the necessary information, or lead to an infinite regress of reconstructions of the present by way of reconstructions of the past.[57]

The optimization reconstruction can be meaningful as a heuristic principle for the reconstruction of partial models, if it be kept in mind that it is a model. „Economics of surviving, placed along with the economics of acting and of thinking, means ... an interpretation of the process of selection as a continuing betting on those organic features that have comparatively great probability to survive in a given environment. A tautology seems to have provided guidance to most fruitful descriptive research. ‚What has a high probability to survive will survive, with high probability', is a tautology. But it has directed creative attention to the question: why in terms of known physics and chemistry, ... has such-and-such feature of an organism a higher probability to survive in such-and-such environnent, than this-and-this other feature".[58]

---

56 Cf. ALEXANDER (1974), p. 374: „What is it, after all, that selection is maximizing".

57 This constitues an inadmissable totalization and universalization of the empirical model of research, making it a speculative principle of totality. According to KRINGS (1984), this transition from a partial empirical theory to a universal principle, this *pars pro toto, is* a logical error of the type METABASIS EIS ALLO GENOS; it designates the point where empirical evolution research is transformed into evolution metaphysics and where scientific theory encroaches without reflection upon philosophical terrain.

58 MARSHAK (1974), p. 380.

If, however, the total process of evolution is understood as optimizing adaptation, this presupposes the transcendental a priori, prior to all experience and empirical verification, that adaptation leads to optima. It presupposes a sort of purposiveness of nature embodying a teleologism that goes beyond the teleology of the Aristotelian ontology of natural kinds. Already in order for adaptation to be conceived as the basic relation of the organism to its environment leading to *local* optima, a continuity and loyalty of reality must be presupposed. Otherwise constancy of observable phenomena and forms, or even species would prove impossible even over limited durations. If adaptation is conceived without continuity of form, this means there exists nothing more than states in a total process of matter, only a flux of the total reality, but no phenomena of trueness to form or distinguishability between organisms and their environment.[59]

The same holds true for the optimization argument as for the Darwinian theory of selection: both refer first and foremost to the relation of the organism to its environment. They both stress the external aspect of the organism's adaptation to given conditions of the outer world -- ceteris paribus -- but can only presuppose, not reconstruct *in entirety*, the inner aspect, the „inner world“ of a species's genesis and the genesis of the setting in which a species or individual exists.[60] As a principle explaining the genesis of the total context of life, the selection principle also makes the transcendental presupposition that the origination of species by selection can be inferred even for the „missing links“ even there where the selection process cannot be reconstructed. From the observable or reconstructable selection processes it concludes that all existence arose as a product of similar selection processes.

This inductive conclusion is possible only with the a priori that all selection acts in the sense of preserving adaptations. Therefore, the neo-Kantians interpreted the theory of selection as an empirical application of the transcendental principle of purposiveness, which Kant defined as „purposiveness of an

59 LEWONTIN (1978), p. 169: „Continuity and quasi-independence are the most fundamental characteristic of the evolutionary process ... Continuity means that small changes in a characteristic must result in only small changes in ecological relations ... Quasi-independence means that there is a great variety of alternative paths by which a given characteristic may change, so that some of them will allow selection to act on the characteristic without altering other characteristics of the organism in a countervailing fashion.“ It follows from Lewontin's critical representation of the adaptation principle that the concept of species cannot be abandoned.

60 Cf. PETERS (1972), p. 349.

in itself contingent connection of multiples". „The natural selection explains in a mechanical way the existence of *purposive* forms. The selection principle assesses all processes from the standpoint of usefulness in the sense of conduciveness to preservation. Therefore, it certainly hits the mark to designate Darwin the ‚greatest teleologist'."[61]

Since reconstructions of the genesis of organisms - in selection theory as well as optimization theory - are possible only as partial reconstructions, but not total reconstructions of the history of nature, it is doubtful that transspecific evolution is explained by selection theory and that the concept of species can be abandoned.[62] In terms of game theory, the theory of selection does name the „operative" rules of play and of the successful game strategies in evolution. But it is questionable whether it „explains" and reconstructs the original endowment and the status-quo-point[63] of the species playing against one another. The question also arises whether, in extrapolating the principle of selection to the whole of reality, the theory of evolution does not overdraw the account of explanations, by writing a check for explaining totality that is covered only for the explanation of partial connections.

3. The Best of all Possible Worlds and the Impossibility of its Total Reconstruction

Sociobiology claims to be able to „reconstruct" a theory of totality by way of the theory of the evolution of social behavior within general evolution theory. Wilson and Lumsden[64] see the explanatory power of a scientific theory in its capability to reduce the greatest possible number of phenomena to simple relationships in an aesthetically satisfactory way. The experience of science had shown that this proceeds best of all by viewing the real world in a matrix of possible worlds.[65] Dawkins adopts this program of possible worlds by reconstructing his evolutionarily stable strategies with the model

---

61 FELLMANN (1977), p. 294.

62 According to LEWONTIN (1978), p. 159, adaptation does not improve the organism's chances for survival, but only compensates for changes in the environment, and makes the organism keep pace with these changes. Transspecific evolution, however, can hardly be explained by adaptation as keeping pace with environmental change.

63 Cf. LUCE/RAIFFA (1957), p. 124ff.

64 WILSON (1978), p. 18; LUMSDEN/WILSON (1981), p. 346.

65 LUMSDEN/WILSON (1981), p. 2.

of the best of all possible ways of behavior. Two objections can be raised against a monistic theory which views the whole of reality as the result of a unified evolutionary process and reconstructs it in a model of possible worlds.

First, it makes no sense to speak of possible worlds when the explanation demanded concerns a process by definition unique, i.e. evolution as the history of the universe. The answer to the question how this unique process could be reconstructed cannot appeal to possible worlds and simulations of developmental paths, i.e., partial developments, because these are precisely not identical with the object of our interest. Eigen and Winkler[66] concede that „only selected original events can be experimentally tested, but not the historical chain of events of evolution". The real world's development in entirety ultimately remains inexplicable in all causalistic, optimization and game theoretical explanations. Fixing antecedent conditions for models, which excludes irrelevancies or provisionally presupposes a point of departure, is inadmissible for the investigation of the whole and of the totality of the process. If the total system of the world is to be explained, no „environment" can be excluded as irrelevant. On the other hand, the attempt to gather up the antecedent conditions and to define the system as a whole leads into an infinite regress.

Secondly, to begin with possible or even the best of all possible worlds as a heuristic model makes sense only for partial worlds with reference to other partial worlds, but not for total evolution. World and evolution can be contingent and possible, and evolution could have proceeded otherwise only for a divine consciousness; for a finite consciousness they are, as wholes, just as they are. Only of partial worlds is a finite consciousness able to say that they are optimally adapted to such and such purposes. Only an absolute and infinite consciousness would be able to say that the world is the best of various thinkable or possible ones. Applied to evolution as a result in entirety of optimizing genes, the optimization argument leads either to a superteleologism of genes as eternal final causes or to a theodicy which interprets evolution as the development of the best of all possible worlds.[67] Both positions are incompatible with a causal-mechanical theory of evolution.

---

66 EIGEN/WINKLER (1975), p. 195.

67 Cf. also GHISELIN's (1974), p. 131, criticism of H. R. Walace's belief in harmony, that natural selection always prevents the evolution of non-adaptive traits.

The problem of reconstructing sociobiological behavior with the aid of possible optimal strategies demonstrates once more the incompatibility of arguments from a theory of selection with arguments from a theory of optimization, as long as no theorem has been proved which shows that selection leads to optimality or that optimality is selected. As-if-explanations and reconstructions in accord with optimization arguments, such as in Dawkins, are not explanations in a strict sense, but rather are plausible stories about what genes should do if they want to secure their survival. As plausible stories, they possess a heuristic value. The difficulties with a „total reconstruction" show that a theory of the whole of reality cannot be conceived by an extrapolation from partial models, but rather that other criteria for truth and other thought procedures, namely those of metaphysical thinking, apply and be taken into consideration.

Chapter IV

Natural Economy Versus Social Economy

The second chapter „sociobiology as bioeconomics" showed how biology and economics converge in their attempt to achieve a unified descriptive and normative explanation of behavior. Gene survival and the maximization of inclusive genetic fitness are understood in sociobiology and bioeconomics as the uniform, final purpose in the behavior of all living beings. Yet it became clear that the sociobiologists do not apply unrestrictedly the „categorical imperative" of sociobiology „Maximize your inclusive genetic fitness!". Not all actions that serve this purpose, not all gene maximizing strategies, e.g. cannibalism, are approved or accepted.

1. Being and Ought, Explanation and Justification

These limitations to the validity of the genetic imperative show that the difference between explanation and justification also arises in sociobiology, and that biological monism by itself cannot be held up. The difference between being and ought, between biological and ethical argumentation arises in the theory itself and thus raises doubts about the theory's monism. The explanation of the evolutionary process is not identical with its justification.[68] We cannot conclude from an understanding of evolution that it also ought to be so, when another course of evolutionary developments is possible and is subject to human influence. Yet all sociobiological authors are convinced of the possibility of redirecting the evolutionary process once we have discerned it, although this possibility of reflection about and redirection of evolution, which presupposes a sort of freedom of theory, is difficult to conceive of in a strictly deterministic model.

How can it be explained that, according to Wilson, the human species has the ability to alter its own nature?[69] Why should linguistic communication free man from the restraints imposed by Darwinian principles, if the laws of evolution are strictly valid?[70] Freedom and self-transformation in accord with self-posited purposes make no sense in a strictly deterministic or stochastic system. What it means to say that people can have knowledge of themselves and of the process in which they and their environment are embedded, cannot be grasped deterministically, because such knowledge presupposes and implies a distance and independence from this causal determination.

The introduction of theory into the system, the doubling of reality by its conceptual representation and negation, and the effects of this image of reality upon the shaping of reality cannot be adequatly understood by mechanistic theories of selection. A determination of the relationship of the mental sphere to the sphere of nature requires, as will be demonstrated, a more complex theory than that of causal-mechanical evolution. The relation of theory and reality, of knowledge and practice, shows that a strictly monistic theory of the whole of reality is possible only as a theory of absolute mind, as in Hegel, not as a deterministic-mechanical one. Monism is conceivable only as monism of mind, in which nature is conceived as the other of mind, and mind is speculatively conceived as that which reaches over the opposition between nature and thought.[71] Deterministic-mechanical theories, on the contrary, always necessarily fall into the Cartesian dualism of „res extensa" and „res cogitans", matter and thought, rendering a consistent monism impossible.

According to sociobiology, as well, ethics is not completely reducible to biology. The principles of ethics cannot be equated with adaptation and opti-

---

68 Cf. NAGEL (1980), SINGER (1982), VOSSENKUHL (1983), WICKLER (1983).

69 WILSON (1978), p. 196.

70 So EIGEN/WINKLER (1975), p. 288.

71 Cf. on Hegel's monism of absolute spirit HENRICH (1980).

mization principles; the principles for guiding cultural-social evolution cannot be equated with the principles of natural evolution. With a view to socioeconomic evolution, the matter can also be stated thus: the social strategies and institutions that arose by way of social and biological evolution and selection need not necessarily be the ones that bring about a social optimum and the greatest unfolding of human potential. Evolutionary developments can likewise lead to socially undesirable conditions and to social dilemmas.[72]

2. The Prisoners' Dilemma and Rational Foresight as the Specific Difference of Social and Natural Economy

A typical such dilemma is the situation depicted by the prisoners' dilemma. The result of social interaction that occurs by social cooperation and when all or some anticipate the general good is more advantageous for all than when all agents pursue only what first strikes them as their own narrowly defined interests.

The prisoners' dilemma is illustrated by the following story: Two prisoners who will appear in court are known to have committed a major crime together. Nevertheless, the court does not have evidence sufficient to convict them for this crime, but only for a lesser crime. The judge asks both prisoners separately if they are willing to confess. If both confess, they will be punished for the greater crime, but will receive a lesser sentence, 10 years in jail. If neither of the two confesses, they will only be punished for the lesser crime and receive two years. If one alone confesses, he will go free as the chief witness and the other will receive the full sentence of 20 years. Faced with the choice, each of the two will argue that it is better for him to confess if the other confesses, and not to confess if the other also does not. Therefore, both will decide to make a confession. Thus, both prisoners go to jail for 10 years, although both would have gotten away with 2 years, if both refused to confess.[73] The prisoners' dilemma is a form of fallacies of composition in individualistically oriented, evolutionarily developing contexts of interaction.

The fallacy of composition consists in wrongly concluding that what is advantageous for each as an individual is also advantageous for the community as a whole, for all together. The interaction of individual strategies,

---

72 Cf. BUCHANAN (1975), p. 167; VANBERG (1983).

73 On the Prisoners' dilemma cf. RAPOPORT/CHAMMAH (1956) and KOSLOWSKI (1982), pp. 7 and 172.

when each does what is advantageous for himself, can evolutionarily result in a state that is less advantageous than what would have occurred if the individuals had anticipated the total result. The cause of such a suboptimal solution resulting from the evolution of individualistic behavior patterns and the structures „composed" of them is the following: Side and reciprocal affects of individual decisions and actions on others and on the subsequent total result, just as the effects from the aggregation of individual strategies, are not or cannot be foreseen and considered by the individuals. There is a summation problem. In evolutionary contexts, numerous individually negligible side effects add up to a weighty undesired total effect. In such situations, individual anticipation of the total result, or anticipation by a central coordinating authority, is able to reduce the uncertainty about and damage from interactions and interrelations of the total process.

Human reason is the faculty of anticipating the universal and of drawing conclusions. Individual and social learning by communication and mutual agreement can avoid or mitigate fallacies of composition, situations like the prisoners' dilemma, and evolutionary misdevelopments.[74] The human ability to anticipate complex effects and interactions, and the general interest in individual decisions distinguishes human action from animal behavior. Anticipation, generalization, and role reversal determine human action in addition to the immediate pursuit of individual interest. At the same time, these qualities constitute the essence of morality and cause the inapplicability of any sociobiological synthesis that denies the existence of the human moral faculty. The faculty of learning always implies an ethos of wanting to learn, a decision. Man learns not only out of necessity; he also learns because he *anticipates* necessity. He has the ability to avoid the situation of the prisoners' dilemma by rational foresight.

Game theory situations such as the prisoners' dilemma also occur in nonhuman nature. The skimming-off principle expresses such a dilemma. In all organisms, the balance between natality and mortality is, due to lack of foresight, effected by the skimming-off principle: more offspring are produced than can be provided for in the environment and the surplus is reduced by selection. Overpopulation is not anticipated and reproductive behavior not preventatively reduced. In consequence, any reduction in mortality causes the population to move over the bearing capacity. The consequences can bring overuse of resources, environmental destruction, and therefore a long-term sinking of the bearing capacity, which is then below that which would be at-

74 Cf. KOSLOWSKI (1983a), p. 174ff.

tained by learning.[75] Because situations like the prisoners' dilemma among animals and plants are not resolved by anticipation and learning, but only by genetic selection, they lead to suboptimal solutions in the interactions within one species and between different species for the period in which they occur. Since genetic selection works much slower than Lamarckian selection by learning, this „selection failure" results in an economically sub-optimal use of resources by way of overuse and non-cooperative competition in situations like the prisoners' dilemma.

A problem of overuse of the Commons arises, for example, in the use of a territory by various species. For the same reason, breeders, farmers, and shepards can increase the returns from nature's capital by „rational selection": they can transform biological externalities between species into mutual economic advantages for themselves and for these species. Since man is able to bypass the skimming-off principle by rational foresight, human knowledge brings a new element into the evolutionary process, defeating evolutionary monism. Man is able to improve or even „perfect" nature. „Nature unaided does not reach an optimum."[76]

Rational foresight is also the reason why all equations of market competition with natural selection are poor analogies.[77] The selection of suppliers according to market-success in the social economy is fundamentally different from that of individuals or species in the natural economy, because the suppliers and demanders not only act and react, but, in their actions, also anticipate the expectations and reactions of the other interacting party. Expectations or anticipations of anticipations are formed. The individuals' anticipations enter into their perceived „image" of the environmental situation and thereby into their strategies.[78] Entrepreneurial success in the market is determined essentially by this ability to anticipate expectations and to make correct predictions amid conditions of genuine uncertainty about future events.

Insofar, Alchian's equating of biological with economic selection as a process of chance selection is misleading for two reasons. First, it overemphasizes the undoubtedly significant factor of chance in the confirmation of successful economic strategies by profit. To make entrepreneurial decisions on strategy by a procedure of chance would certainly be not profit maximizing strategy. Secondly, as was demonstrated with the problem of the skim-

---

75 Cf. MARKL (1980b), p. III.

76 TULLOCK (1971), p. 391.

77 This identification can be found in ALCHIAN (1950) and FRIEDMAN (1981).

78 Cf. KOSLOWSKI (1983b), p. 310.

ming-off principle, natural selection and economic rational selection are distinct. Their difference consists in that social systems learn and can pass on in the same period of time what was learned in the form of communicable knowledge. In contrast to organisms, which are selected, firms and social groups are in a position both to anticipate future developments and to pass on in the same period learned patterns of behavior, insights, and experiences in general to members of the same species. If economic, entrepreneurial, and organisational learning are to be described in categories of evolutionary theory, then this can occur only in Lamarckian categories, which go beyond the Darwinian theory of evolution. Darwinian categories cannot adequately explain evolution by learning.[79]

3. Gene Survival Under Essential Constraints

In general, economic imperatives are valid only under ethical and social constraints,[80] whereby these constraints usually cannot be further justified by efficiency either economically with respect to „utility" or biologically with respect to gene survival. Such a teleological justification with reference to a universal optimization argument is not even possible.[81] A sociobiological justification of ethical constraints which is based on their necessity for the preservation of the gene pool is, in complex situations of social interaction, not realizable in optimization theory, because the information necessary for such a statement is not available. We cannot demonstrate whether the prohibition of cannibalism is or is not a strategy for gene optimization for an individual in a particular society, due to the complexity of the optimization problem.

The reason why such social constraints are not only valid as restrictions on our actual willed action, but also are valid in general is that we do not relate our own and others' life functionally to gene survival; rather, we our-

79 GOWDY (1980) points out that the higher velocity of Lamarckian evolution permits man to excape the limitations of natural selection.

80 Cf. HIRSHLEIFER (1978b), p. 321 and (1984) and BUCHANAN (1982) on the relationship between „natural economy" and „political economy".

81 Cf. the analogous critique by SPAEMANN (1981) on attempts to base ethics on the universal teleology principle ‚an action is good when it maximizes the amount of goods in the world'.

selves desire these restrictions as qualities of human existence.[82] We desire concrete forms of life, not the survival of genetic programs. According to sociobiology we live behind a veil of ignorance about our actual purposes. What we want, namely, a good life, is the illusive manifestation of a hidden purpose: the optimization of the survival of genes.

Now it might be asked what argument could enlighten us as to our false consciousness and its idealism. Economizing genes at work are not observable, and the Cartesian doubt, which draws everything into doubt, leaves us only with one certainty: that *I* think, not a gene. The sociobiological argument about the actual content of our willing can be grounded neither empirically nor by introspection. In every way it is transcendent and metaphysical. Population statistics do not confirm it either: wealthy societies do not expand their reproductive behavior to production possibility frontier of gene production. Their populations grow slower than those of poorer societies. They do not maximize gene survival, but rather a utility function, the argument of which encompass more than mere gene survival. Thus, as concerns populations of advanced societies which no longer live on the borderline of subsistence, the assertion of gene maximization is an anachronism.

The response of bio-economics to this empirical observation is that people in advanced societies maximize not the quantity but the quality of children. But what does it mean to maximize the quality of children? Does it mean maximizing survival machines made by and for genes? A machine that is to produce gene survival will be measured on the amount of the desired output, and not on the quality of the input for this output, here the children. Maximizing quality instead of quantity of children means precisely that more arguments enter into the goal function of production decisions than mere gene survival. The quality of children who can live a good life is incommensurable with the quality of survival machines. However, societies in which the economic pressures of the struggle for survival are alleviated behave contrary to the prediction of sociobiology. The theory of gene maximization will, therefore, have to be considered refuted.

All „laws" of motivation psychology and demand theory contradict sociobiology. According to Maslow's pyramid of needs, upon increasing satisfac-

82 Cf. McCain (1980), p. 127: „Economists' models are models of *constrained* optimization. The sociobiologists' models are not ... This does not mean that the sociobiologists' models are simpler but that the sociobiologists often overlook constraints on maximization. Constraints are commonly important. Constraints may well be more important than maximization itself."

tion of physiological needs the urge toward higher, spiritual and social needs grows, which become increasingly distant from gene maximization. Likewise, people with growing income demand superior goods farther removed from reproduction, while expenditures for reproductive means diminish (Engel's law). The citizens of West Germany, for example, spent more in 1981 for hobbies than for food.

This development from physiologically to culturally determined needs confirms that culture and its experience of meaning belong to the original needs of the human condition. Man cannot pass his life maximizing the survival of genes, because this would not have enough meaning for „man - the bored animal" (W. Sombart). Humankind disposes of a surplus of meaning in its endeavor to understand the world and master its life, a surplus that could not be satisfied by biological functions. „The world never has enough meaning ... the mind always has at its disposal too many connotations for the objects".[83]

It is difficult to reconcile these sociologically observable phenomena and developments with a hypothesis that is immunized against all empirical arguments. In view of the evidence from immediate experience in our every-day life, in our *„Lebenswelt"*, and from sociological observation, an astonishing „materialistic idealism" is required in order to hold onto the materialistic hypotheses and assert an ontological primacy of gene maximization over practical self-experience. Self-experience knows the desire for reproduction, but not for maximal reproduction.

It could be maintained that the difference between the reproduction drive and gene maximization is not so considerable. But precisely the maximization hypothesis renders the sociobiological synthesis possible as a unity of sociology, biology, and economics, because it alone can justify the theory's great scope covering many domains of life, and it alone would constitute the theory's predictive power. As soon as reproduction is only *one* argument in the individual's maximization of his utility function over his life-time, the *unity* of economics and biology becomes untenable. Biological purposes then become purposes among other purposes, which must be efficiently fulfilled in every social order. They are no longer the common denominator of all other purposes.

The differences cited between social and natural economy show that a direct transfer of biological, Darwinian categories to the social domain is inadmissible. They show that sociobiology, if its synthesis is conceived as a

---

83 LEVI-STRAUSS (1981), p. 202.

neo-Darwinian monism of maximizing gene survival, will suffer the same fate as social Darwinism. Like the latter, it will be a theory with mere ideological significance.[84] Yet sociobiology, like social Darwinism, also cannot be regarded as the ideology of a market economy. The order of the market economy is distinguished from the objectivism and monism of sociobiology by its pluralism of values and subjectivism of needs. The economic order should not fulfill predetermined or imposed purposes, but rather the subjective, freely-chosen needs of individuals. The economic order should make freedom possible, i.e., permit individuals to set their subjective goals themselves and pursue them in compulsion-free coordination with others' plans.[85]

As against a social-Darwinian objectivism, advocates of economic liberalism like Hayek and Knight emphasize the cultural plasticity of human needs, subjectivism in value theory, and the significance of freedom in social and economic evolution.[86] Since economic liberalism stresses the significance of the market as a process of discovery, and a process of social and individual learning, it always understood its evolutionism to be Lamarckian, not Darwinian.[87]

4. Sociobiology and Natural Right

Whether the reproductive imperative is viewed as a categorical imperative to *maximize* gene survival, or rather sensibly as *one* imperative among others, decides whether a scientifically fruitful or an ideological form of sociobiology is propounded. In recalling the biological basis of humankind and society, sociobiology is able to fulfill an important task for the social sciences. If the hypothesis of gene maximization is reduced to the weaker rational hypothesis that human beings want to reproduce themselves and that this will is genetically guided, but that not all social purposes are teleologically related to gene survival, then sociobiology can make a fruitful contribution to the idea of a philosophical natural right.

The endeavor to synthesize biology and sociology corresponds to the effort at grounding a natural social order by natural right. Plato and Aristotle

---

84 Cf. on social Darwinism CONRAD-MARTIUS (1955), KOCH (1973), CHERRY (1980), MARTEN (1982).

85 Cf. KOSLOWSKI (1982b).

86 HAYEK (1974), KNIGHT (1935).

87 Cf. FELLMANN (1977), p. 291; PETERS (1972).

see the nature of the human and of human society in a continuum from the plant and animal soul up to the human soul and consider the social order to be teleologically related to providing for the purposes of this continuum.[88] Even if, after the experience of human subjectivity and the autonomy of reason, natural right can no longer be dogmatically presented as a catalogue of norms, it still has today a significant, critical function in recalling the nature of man as opposed to a freedom which asserts itself absolutely, striped of all limits.[89] Natural right is no ideology of early high cultures, meant to lend their contingent social order an appearance of lawful necessity[90], but rather an attempt to reconcile the purposes of freedom and the purposes of the nature of man. The insistence that an order in accordance with human nature is the measure of positive law and of social conventions and norms is not just apologetic but entails also sometimes a high degree social criticism.

Sociobiology can recall and ground the natural basis of human social life by universalisms like reproduction/sexuality, parental care/marriage, territoriality/defense, devision of labor/social stratification. It counteracts the belief of an exaggerated historicism and a hybrid philosophy of history that humankind could fully emancipate itself from these natural conditions by means of society. Sociobiology draws attention to the unity of life and the relationship between human and non-human life.

Also, as against an exaggerated subjectivism in economic value theory, it draws attention to the fact that there are economic purposes given by man's natural basis, which are not a matter of subjective demand and conventional needs. Sociobiology as natural right implies the imperative that political and economic freedom of decision be put in a mindful relation to and be measured on the natural conditions of humankind and its world. Such a sociobiology, as a theory of the fundamental kinship of life forms, requires a mindful relationship of freedom to both human and non-human nature, and it grounds a consciousness of the ecological context on the sociological level.

As a reminder of biological universalisms and genetic restrictions on human plasticity, sociobiology can form a counterpart against abstract historicism and culturalism. In this sense, it would be a theory of human history as a continuation and completion of natural history, but not an identification of the two. Nevertheless, in order to avoid falling itself into an abstract biolo-

---

88 PLATO: *Republic* 466d, 372d-e, et passim; ARISTOTLE: *Nicomachean Ethics* I 6, 1097b 30ff.

89 Cf. SPAEMANN (1977a).

90 So TOPITSCH (1962).

gism, it would have to acknowledge that the „natural" purposes are not the human's definitive purposes. A sociobiology in accord with natural right would have to recognize the reality of human subjectivity and freedom in society and see that it not be sacrificed to biological universalisms. In connection with Aristotelian natural right, sociobiology would have to render it evident that the *ergon tou anthropou*[91], the work of the human, is not fulfilled by the satisfaction of necessary biological conditions or even by the maximization of gene survival, but rather consists in acting in accord with reason.

The realization of humankind's rational nature must include the realization *both* of biological nature *and* of subjectivity. In this regard, sociobiology could, like natural right, check a tendency of the modern consciousness to break apart into a mindless materialism of the mechanical world view and a bodiless idealism that can no longer find a societal and biological basis.

Chapter V

The Sphere of the Mind, or: Unsolved Problems of Sociobiology

1. Intentionality and Theoretical Freedom

Human action in accord with reason can be described by way of intentionality as the ability to act, by way of conceptual thought as the ability to know, and by way of the consciousness of this activity, i.e. consciousness of self. According to natural right, man's self-realization consists essentially in actualizing these faculties in rational politico-social practice and in free, theoretical knowledge. These two human activities, practice and theory, are given in self-experience. We experience ourselves as beings that can act, not merely react, and as beings that can know, not merely perceive. A complete synthesis of biology and sociology, of genetic and cultural learning, would have to be able to explain these phenomena of intentionality, conceptual knowledge, and consciousness of the self as products of *one* evolutionary process, and to understand the phenomena as they are in themselves.

---

91 ARISTOTLE: *Nic. Eth.* I 6, 1098a, 12ff.

Wilson admits that intentionality is still the real secret of neurobiology.[92] As a solution, he suggests viewing the will as the result of a struggle between different wiring schemes. Thereby, spontaneity would be banished from the model. Wilson concedes that the problem of the will has not been deterministically solved, but maintains it is solvable in principle. Lumsden and Wilson consider conceptual thought the decisive difference from animal life. In their attempt to concretely elaborate the synthesis of cultural and biological coevolution, of social and genetic learning, they grant that the conceptual representation of reality and the symbolization of perceived content -- they call it „reification" or „ideation" -- fundamentally distinguish human language from animal communicative behavior.[93]

Lumsden and Wilson view this process as having taken place over the three million years from australopithecus to the late paleolithic age. They explain the acceleration of this process by way of a gene-culture transmission, wherein the passing on of „culturgens", of cultural information bits is thought to have accelerated natural selection by cultural learning and therefore said to explain the greater speed of cultural evolution. The transmission of cultural learning follows epigenetic rules for forwarding information in our thought apparatus, rules which furthermore are genetically selected as being expedient for reproductive success.

According to Lumsden and Wilson, only the conceptual cognition of a thing as itself, the reification, is inexplicable in the theory of selection.[94] Its conduciveness to survival is not genetically reconstructible. Trivers therefore declares: „The conventional view that natural selection favors nervous systems which produce ever more accurate images of the world must be a very naïve view of mental evolution."[95] Rather, selection must lead to a certain degree of self-deception, since deception is a basic characteristic of communication among animals. Selection must cause a self-deception that permits some facts and motives to remain unconscious, in order that they not betray the practiced deception by way of the nearly imperceptible signs of the agent's awareness of his deception.[96]

---

92 WILSON (1978), p. 76.

93 LUMSDEN/WILSON (1981), pp. 5f. Cf. the review of this book in MARKL (1982).

94 LUMSDEN/WILSON (1981), p. 346.

95 R. L. TRIVERS in the introduction to DAWKINS (1976), p. VIII.

96 *Ibid.*

Conceptual knowledge is inexplicable in sociobiology and, above all, dysfunctional for a survival strategy. This is confirmed by philosophical tradition, which regards conceptual thought as an expression of theoretical freedom, as a „free self-disclosure and opening" to the world,[97] and as the result of an exoneration from action. Its high value is thought to lie not in its survival utility for the individual, but rather in a knowledge having purpose and value in itself, a knowledge of that which in truth is.

Another notion may be added which has been held from Plato to Hegel, that the highest form of theory turns into the most useful practice, because it discloses Being and the Good to us, and therefore renders our politico-social reality transparent and makes it possible for us to establish a good order. Such knowledge accomplishes more than pure survival knowledge: it enables us to know what the good life is -- individually and politically -- and to expand the range of social action from that of mere survival to that of a good and rewarding life.

The purpose of conceptual thought does not consist in merely securing the reproduction of genes, but rather in opening up a space for the self-realization of the human as a rational being. This enabling of the realization of reason by reason is, according to Plato in the *Republic*, at the same time both a purpose in itself as activity of the theoretical *logos* and a condition for meaningful politico-social practice as practical political *logos*.[98] Thinking itself is both purpose and enjoyment, and it is simultaneously a capacity for pushing back the frontier imposed by the death of the human individual. It possesses both a value in itself, and a sociobiological functional value. Knowledge of the structure of the real as real and the Good as the Good, not as something merely useful for survival, makes it possible -- individually by way of a proper life and politically by way of a political order arranged according to this knowledge -- to delay death.

2. Consciousness of Death, Consciousness of the Self and the Choice Between Death-delaying Strategies

This leads to the third unsolved problem of sociobiology: consciousness. Consciousness is above all consciousness of death. With reference to Husserl and Derrida, Wheeler shows to what extent consciousness of one's own iden-

97 KRINGS (1966), p. 37.

98 PLATO: *Republic* 473c.

tity is connected with consciousness of the possibility of being-no-more, of death, and how it shapes our life in contrast to the life of animals, which possess no consciousness of their finitude and identity.[99] From the viewpoint of gene maximization, awareness of self and of death are detrimental. The individual ought to understand herself as a transit station for eternal genes and not hang on her own fleeting life.

Awareness of death is also harmful to a being who must die for certain and whose securement of life is always only a delay of death. Thus, consciousness of death and self-knowledge are the expression of an evolutionary banishment from the paradise of blind, programed gene maximization. According to Plato, the awareness of death is the beginning of human concerns and the beginning of thought. Death stands at the beginning of a movement of thought, which attempts to recognize Being and the Good, because the inquiry into the proper life arises only upon knowledge of mortality and with the possibility of missing the meaning of life.[100]

Intentionality, theoretical knowledge, and consciousness refer to one another. Practical and theoretical freedom in the consciousness of subjectivity are responses to the knowledge of death, because death tears open the perspective of care, responsibility, and justification of one's own life. Against this background of delaying one's own death and justifying one's own life, the capability to acquire an increasingly precise conceptual image of reality and act accordingly gains - contrary to Trivers - decisive significance. Reason comes to be the precondition of death-delaying and life-fulfilling strategies for the individual and society.

Practical freedom (intentionality) makes possible the transcendence of immediate animal life and self-interest; theoretical freedom for truth makes possible knowledge of the Real as the Real, and thereby knowledge of the best sociobiological order for man. Consciousness of death is the price and precondition of these „freedoms". Plato systematically depicted this connection for the first time in the *Republic*. His kingship of philosophers is a sort of correct sociobiology, which also considers its own application in the theory and views itself as part of a strategy for postponement of death. Adequate knowledge of reality is a precondition for a rational social order, which goes beyond a republic of animals.

The separation of humankind from the immediate biological life by intentionality and freedom of knowledge is the lance that wounds *and* heals. The

---

99 WHEELER (1978).

100 PLATO: *Republic* 328c -331b, and the concluding myth.

distinction of the human from his or her socio-biological function by consciousness of the self and intentionality effects the disunion of individual and society, selfishness and altruism. Yet the theoretical and practical freedom that follows from this disjunction is also the ability to recognize and do the Good and to overcome the disjunction.

3. Being-with-oneself-in-another as Reconciliation Between Self-interest and General Good

Consciousness of the identity and non-identity with oneself is the precondition for altruism. To a selected, natural system without consciousness, everything always is just as it is. It has *no consciousness of negativity*, that a state of existence is other than it could or ought to be.[101] But it is true of human consciousness that it is a consciousness of the „other than itself“ that it can be aware that it is different from itself; that it is not fully itself. It can also know that another is also a self, that he or she possesses self consciousness. The human can be with herself in another person. Hegel called being-with-oneself-in-another both „love“ as well as „freedom“.[102] Consciousness, freedom, and love refer to one another. This triad also transcends the mechanistic interpretation of social life. At the same time, it refutes reductionist descriptions of the altruism problem in sociobiology.

---

101 Cf. SPAEMANN/LÖW (1981), p. 251.

102 HEGEL: *Wissenschaft der Logik* (Science of Logic), Part II, Chapter 1, A: „The universal (of the concept) is therefore the *free* force; it is itself and it overlaps its other; but not as something *violent*, but rather that which calmly rests the other and is *with itself*. Just as it has been called the free force, so too it could be called the *free love* and *boundless joy*, for it is a conduct toward the *different* only as *toward itself*; being in the other, it has returned back to itself.“ See also Hegel: *Philosophie des Rechts* (Philosophy of Right), § 158 (Addition): „Love means in general the consciousness of my unity with another, so that I am not isolated for myself, but rather gain my self-awareness only by abandoning my being-for-myself and by the self-knowing as my unity with the other and the other's with me ... Love is therefore the greatest contradiction, which the understanding cannot resolve, in that there is nothing harder than this pointedness of the self-awareness, which is negated and which I should still affirmatively possess. Love is the simultaneous generation and reconciliation of this contradiction.“

Altruistic behavior is a contradiction for a theory that assumes the thorough determination of behavior by selfish genes. As shown above, the solution generally consists in maintaining that the supposedly altruistic behavior, e.g. of parents, is simply an extended selfishness. „Our virtues are only concealed vices" (La Rochefoucauld). It makes no difference to this encompassing genetic selfishness whether 100 % of its own genes survive when it does not help its children, or 2 times 50 % of its own genes survive if it dies while helping but two of its children live on. Such altruism is a somewhat expanded programed selfishness for maximizing inclusive genetic fitness. Sociobiology represents altruism in an exaggerated fashion as a self-sacrificing for others. Having thus defined altruism, it is then shown that such exaggerations do not or cannot exist, or that sacrifice-altruism is really no sacrifice, because the sacrificer and the sacrified are not at all identical in the self-sacrifice of the altruistic individual. For the agent does not sacrifice himself; rather, genes sacrifice him for *their* duplicates in *his* offspring.

The representation of ethical behavior as „sacrifice" contradicts the purpose and grounding of ethics. On the contrary, ethics calls for overcoming the difference between self-interest and interest of the whole.[103] It calls for viewing the other as oneself; it does not call for sacrificing oneself for the other. Transcendence of subjectivity presupposes freedom, love, and consciousness, it presupposes being-with-oneself-in-another. Ethics is possible only by way of consciousness, freedom, and love. The ethical transcendence of subjectivity enables man to decide between different forms of social organization and to freely acknowledge a good order.

Consciousness and freedom make possible morality as the overcoming of immediate self-interest for the sake of a common weal, which is both in one's proper interest and in the interest of others. The conscious individual is able to choose among different universalizable, death-delaying strategies and to anticipate their potential for delaying death. The disjunction between I and society, between identity and non-identity of the I and the others, can be reconciled through a free, loving consciousness, whereby the individual is with himself in another, is simultaneously himself and the other.

---

103 Cf. KOSLOWSKI (1982), pp. 67ff. - TRIVERS's (1971) analysis of altruistic behavior is more differentiated and does more justice to the multiple intertwinings between self-interest and general interest, and also forms of hypocrisy and „moral aggression" than does the analysis of some sociobiological economists.

The dialectic of division and reconciliation in the triad consciousness, freedom, and love explodes the causal-mechanical model of genetic sociobiology, because the consciousness is itself and everything else at the same time. „The soul of man is in a certain respect everything *(quodammodo omnia)*" says Aristotle.[104] Consciousness is dialectical in itself and therefore incomprehensible in a crude determinism. The reason why sociobiology, as a causal-mechanical theory, cannot explain the evolution up to conceptual thought is due not lastly to the dialectical nature of consciousness. A sociobiological synthesis, however, which does not take consciousness, freedom, and knowledge into account in its theory of the Social because they are deterministically inexplicable, is not a synthesis, but a contraction of social reality, because the phenomenological evidence from these elements of human life on the basis of self-experience is stronger than the evidence of an incomplete causal-mechanistical model.

Chapter VI

Sociobiology as a World-view and the Choice Between World-views

1. Sociobiology's Claim to Truth as a Problem of Theoretical and Practical Certitude

The elaboration of a genuine socio-biological synthesis is itself an ethical problem of theoretical and practical freedom. The problem of theoretical freedom consists in the question -- proper to all scientific research -- when a hypothesis can be freely accepted and considered sufficiently grounded so that it can claim social validity. For Ramsey, the problem of induction and inductive confirmation is a practical problem that in essence can only be solved probabilistically. For him, it consists in the question, at what point is a researcher prepared to bet on the correctness of his hypothesis.[105] All induction is a problem of probabilistic, practical certitude. This applies especially to

104 ARISTOTLE: *De anima* 430a.

105 RAMSEY (1926). On the consequent necessity for an „ethos of thought" cf. KRINGS (1956), p. 29.

the test of a hypothesis that is so comprehensive and so overturns our conception of the world and ethical views such as sociobiology.

The question is: what degree of justification can the hypothesis of cultural-genetic coevolution claim as a materialistic-deterministic theory, if central phenomenal domains such as consciousness and conceptual thought cannot be remotely explained within the model or are even denied. To designate the graphical and mathematical depiction of self-reenforcing feedback loops a fitting model for explaining the evolution of simple neural processes of sensation into conceptual processes of cognition, as do Lumsden and Wilson, is an unproven claim. Faced with such a claim, philosophical criticism must adopt the role of one who insists on the verification of these claims. In a social theory of the scope of sociobiology, the anticipation of research results that will *perhaps* be obtained sometime, and treating them as already proven, is justifiable neither by philosophy of science nor by the ethics of theoretical research.

It is a peculiarity of modern science since Descartes to constantly anticipate its own fulfillment and completion -- which is supposed to consist in the complete transparency of the world for human knowledge -- at a time when this transparency of reality is still out of the question. With its claim to total explanation, sociobiology is also in the situation of someone who promises more than he can fulfill. Descartes had formulated a program of mechanistically explaining the world and society in a model of totality that would culminate in medicine. Once founded upon a complete knowledge of the biological and social nature of humankind, it would be able to realize the human dream of happiness and eternal health in a comprehensive, scientifically founded, medico-political praxis. According to Descartes, the total science will replace insufficiently grounded, insecure, and only morally certain ethics after completion of a centuries-long process of research.

This dream of scientific total transparency and the resulting praxis also guides Wilson: „To chart our destiny means that we must shift ... to precise steering based on biological knowledge ... Above all ... ethical philosophy must not be left in the hands of the merely wise."[106] Wisdom is to be replaced by science, the insights of ethics and politics by the insights of sociobiology. Finally, the position of political wisdom in guiding society is to be taken by prevision control of human and societal evolution along the lines of sociobiological theory. However, since the instruments of sociobiology are

106 WILSON (1978), pp. 6f.

still poorly proven hypotheses and models, this amounts to wanting to steer a tanker with a paddle.

Descartes was more self-critical on this point. As long as the program of the total transparency of reality is still a program, we must, according to him, -- *faute de mieux* -- resort to the moral certainty of rationally founded ethics, to wisdom, because this can be replaced only by *complete* knowledge, by total science.[107] „Moral certainty" is that certainty of having judged and acted correctly that is attainable after a conscientious and rational search for knowledge of the situation, after ethical pondering and responsible decision.[108] Since the complete knowledge of total science is not yet available, we must strive for wisdom and moral certainty, we must, like someone lost in the woods, follow simple rules of thumb and orient ourselves upon a provisional ethics of rational consideration and action. Likewise, as long as the truth of sociobiology's materialistic world view and the teleology of gene maximization has not been proven, the wisdom of political and ethical reason must be followed.

Sociobiology claims to expose the ethos of practical reason and every-day communication as a false appearance that conceals the selfish gene. For this exposing, *sociobiology* bears the burden of proof. It must furnish proof of its theory's correctness, because, in the anthropomorphic language of reason, the ethical consciousness can show that it promotes the good. Ethical wisdom does not equate the word „good" with „conducive to gene survival" and it recognizes the validity of the reproductive goal only within the totality of moral and biological purposes, but not as a solitary „final purpose". If we understand ourselves to be rational beings, then this wisdom must take primacy over sociobiological self-programing for gene maximization. Since the sociobiological explanation of reality, which supports its ethics, is incomplete, the theoretical status of rational ethics remains stronger than that of sociobiological gene ethics. We live now and must decide now about social strategies for delaying death and for the good life. We cannot wait in the woods where we are lost until, in decades or centuries to come, a total theory of society instructs us about the structure of the woods and why we lost our way in it.

---

107 DESCARTES: *Discourse on Method*, Part III. Cf. SPAEMANN (1977b) on the relation between Descartes' scientistic program and provisory morals.

108 Cf. H. VAN LEEUWEN: „Certainty in Seventeenth-Century thought", in: PH. WIENER (ed.): *Dictionary of the History of Ideas*, New York (Scribner's) ^{2}1973, vol. 1, pp. 307-9.

2. Sociobiology as a Myth

As a theory of totality, sociobiology has the extra-scientific character of a world-view. As theoretical evolutionary monism, it is *one* form of *prima philosophia*, one form of metaphysics. Its arguments, insofar as they relate to the entire process of the universe, cannot be confirmed any more than refuted. Wilson concedes that, as a total theory about the coevolution of nature and society from the big bang to modern industrial society, the theory of evolution is not a scientific hypothesis, but an epic, a story. „The evolutionary epic is mythology in the sense that the laws ... are believed but can never be definitely proved to form a cause-and-effect continuum from physics to the social sciences.“[109] However, according to him, „... the evolutionary epic is probably the best myth“[110] we will ever have.

By what criteria do we decide which myth is the best? Does arbitrariness alone rule amid assumptions held, or are there other rational exiteria?[111] For the theoretician Wilson the answer is clear: the epic that *explains* most is the best. The materialism of sociobiology will conquer religion, because it is more successful at explaining theoretical phenomena. Yet the question remains whether this theoretical criterium of explanatory success is the sole criterium for deciding between world-views, or whether practical-social view-

109 WILSON (1978), p. 192.

110 *Ibid.*, p. 201.

111 GRAY/WOLFE (1982) find sociobiology and creationism to be two sides of the same coin. According to them, both are ethno-sociologies and ethno-mythologies which want to remedy the crisis of the industrial society and secular consciousness by way of a new synthesis. According to Gray and Wolfe, neither completely uncouples from the secular world view; rather, they supplement it - sociobiology by „supernature“, creationism by the „supernatural“: „Sociobiology replaces creationism's supernatural with a supernature in which the role of God is played by natural selection ... Creationism finds morality and hope above, sociobiology below reason“ (pp. 585, 589). The authors conclude that both mythologies, sociobiology and creationism, fulfill the same function in American society, namely conceptualizing and supporting existing social institutions and norms. The clever analysis demonstrates the equivalent social function of the two theories of totality which is not surprising, since both are theories of the entirety of reality. The demonstration of function equivalence shows at the same time the weakness of all functional theories: they cannot say which functional equivalent is superior, and why certain function carriers are adopted but not others. There are multiple ways of fulfilling every function.

points must also enter into the decision. Is materialism also more successful at founding a good life in a just society and at answering questions of meaning? With his view to the explanatory function, Wilson overlooks the task of myth and religion in moving man to morality, in helping him to overcome his experiences of contingency, and in presenting a self-interpretation of man in which he can recognize himself as a rational being.

This comes to light in Wilson's reinterpretation of the story of Job: God is supposed to have asked Job for the *explanation* of universal laws of organic and inorganic nature with His question: „Where were you when I laid the foundation of the earth? .... Hast thou entered into the springs of the sea? ... Hast thou perceived the breadth of the earth? Declare if thou knowest at all."[112] Wilson's answer to the book of Job is: „And yes, we *do* know and we have told. Jehovah's challenges have been met and scientists have pressed on to uncover and to solve even greater puzzles."[113] That this was not the question eludes this scientistic braggadocio. Job was not asked to present his scientific explanation, but to show his power: „Doth the hawk fly by thy wisdom ... Doth the eagle mount up at thy command, and make her nest on high?"[114] Job's problem is not a missing explanation, which he doesn't receive either, but rather his experience of impotence and contingency. In the story of Job, God responds to this experience not with scientific explanations, but with the demonstration of the magnificence of nature.

The story of Job makes clear that the theoretical function of religion and world-views is only one, while its practical-moral and emotional mission is equally important. World-views, in which theory and practice unite to form a concept of the good life, not theoretical explanations alone, can fulfill the four tasks of explanation of the world, motivation for moral action, reduction of the human experience of contingency, and human self-interpretation.

3. The Epic of Evolution and Platonic-Christian Metaphysics - a Comparison

In conclusion, the epic of evolution should be compared with the occidental tradition of Platonic-Christian metaphysics according to the criterium of the theoretical *and* practical function of world-views. This comparison sug-

---

112 *Job* 38; 4, 16 and 18

113 WILSON (1978), p. 203.

114 *Job* 39; 26 and 27.

gests itself, because the scientific materialism of sociobiology aims at the unity of theory and practice in aspiring to a precise political control of our evolutionary fate. This comparison is to be joined with the question, what the appearance at this point in time of a new myth such as Wilson's epic of evolution signifies for evolution and the history of the species.

According to Wilson the following statements form the core of the epic of evolution[115]: „... that the laws of the physical sciences are consistent with those of the biological and social sciences and can be linked in chains of causal explanation; that life and mind have a physical basis; that the world as we know it has evolved from earlier worlds obedient to the same laws; and that the visible universe today is everywhere subject to these materialist explanations."

At first, the objection arises that intentionality and consciousness have proved up to the present deterministically-causalistically inexplicable, as Wilson himself concedes, and that the epic is therefore not consistent but self-contradictory. Yet consistency is the decisive criterium for the truth of world-views or theories of totality, because by definition there are no arguments or criteria outside of a theory of totality by which the theory could be measured.

Philosophical metaphysics of the Platonic-Christian sort forms the counterpart to the epic of evolution. Traditionally, it is divided into general metaphysics, *metaphysica generalis*, or ontology, i.e., the theory concerning the basic principles and first causes of being, and into several special metaphysics, *metaphysicae speciales*, i.e., the theories concerning God, freedom, and the immortality of the soul. The matters of these *metaphysicae speciales*, which must above all be at issue here, have made their way in a process of a religious-philosophical „evolution" in the Western world since ancient Greece and Judaism. They comprehend monotheism, the conception of man as a free and a spiritual being, and the view that the soul is immortal and subject to judgment after death.

From monotheism follows the assumption of a consistent and coherent cosmic and societal order in accord with *one* will, not mutually contradictory, polytheistic wills.[116] From the idea of freedom follows the possibility and the demand upon human beings to act responsibly and freely. From the idea of the immortality of the soul follows the demand upon human beings to lead life as a whole responsibly and justifiably. Consistency of the divine and moral world order, freedom, and immortality of the soul form together the

115 WILSON (1978), p. 201.

116 Cf. KOSLOWSKI (1982), pp. 37f. and 95f.

conditions for the possibility of ethics and a consistent life plan, because this is transcendentally grounded and sanctioned by the judgment and immortality of the soul.

Freedom and the immortality of the soul also form together the conditions for the possibility of a rational altruism to overcome immediate selfishness, because love is the expression of freedom in this world view, and love for one's neighbor becomes a meaningful option for life by way of the judgment and immortality of the soul. These cultural and philosophical values are the expression of a „cultural evolution“, of the history of thought in the Western world. The socio-biological functionality of this process of cultural and philosophical-religious development cannot be denied.[117]

Let us compare the elements of the evolutionary epic, which Wilson suggests likewise to be the result of a cultural-genetic coevolution. This evolutionary metaphysics can be summarized as a general metaphysics of evolution (law and chance) and the specific metaphysics of determination of the will and the immortality of genes with elimination of the concept of the soul. The unity of evolution, according to Eigen and Winkler, and Wilson, consists in a concert of chance and laws of nature,[118] while it remains unclear who made these „laws“. Chance becomes a central generative factor, to which man must subject himself in the totality of nature and society as well as history. The theorem of determination of the will makes it difficult to accept ethical norms and renders responsibility virtually ungroundable. The theorem of the immortality of genes justifies a genetic selfishness, which reduces love to an epiphenomenon and makes moral action appear ultimately illusionary. The principle of the interaction of chance rules of play, determination of the will, and selfishness of genes together make it difficult to assume consistent morals and social norms of action as well as a unified rational plan of life.

The metaphysics of sociobiology and evolutionary monism interprete man essentially as a reacting, not an acting being. Thus, as social metaphysics in the sense of a society's ultimate cognitive and normative orientation, the elements of the evolutionary epic do not promote the rational mastery of the world and of life. They are not conducive to rationality in the sense of consistent acting and thinking. From a socio-cultural viewpoint,

---

117 On the logic of the evolution of religions and world views and on the connection of this evolution with the development of ethics and human self-understanding cf. BELLAH (1970), DÖBERT (1973), pp. 256ff., KOHLBERG (1981).

118 EIGEN/WINKLER (1975) and WILSON (1978).

they represent a regression as compared with traditional metaphysics. A genuine sociobiological and social-philosophical-normative synthesis would therefore have to reject the epic of evolution as a societal world-view and final symbolic orientation, as a theory of totality, because it does not conform with society's level of socio-cultural development.

One can dare to predict that the evolutionary epic will not prevail as an „evolutionarily stable strategy" in the competition among world views, because it is not adapted to the demands of the rational society. Wilson must come to terms with the fact, annoying for a thoroughly evolutionary approach, that the selection of world views does not follow the logic of sociobiology, and that the evolutionary epic does not prevail. As he himself writes: „If anything, knowledge is being enthusiastically harnessed to the service of religion. The United States, technologically and scientifically the most sophisticated nation in history, is also the second most religious - after India."[119]

Nevertheless, he does not consider this empirical refutation in the selection of world views to be an objection to materialistic sociobiology, but rather a result of the fact that religion appeals more forcefully to human emotionality. These emotional advantages of religion, however, are not interpreted as functional arguments against the epic of evolution, as the objection that religion might correspond more closely to sociocultural basic human needs and to the evolution of culture. Since the epic of evolution does not apply its own theory of the evolutionarily stable strategy to itself, it is incapable of acknowledging that, as a social theory, it is not in accord with sociocultural evolution and is therefore not an evolutionarily stable strategy in competition with other world views.

Advocates of the evolutionary epic like Wilson contradict their own materialistic-evolutionary approach and are idealists, insofar as they want to hold onto a „scientific materialism" even when the religious world view is selected and proves to be evolutionarily successful and sociologically functional. The contradiction that was noted in connection with Dawkins repeats itself: as a total theory, evolutionism becomes „idealistic" when the totality of its thought is set above the evidence of the phenomena. Because it denies the phenomenon of mind, evolutionism is compelled to conceive a theory of social evolution that can neither include nor explain an essential step of evolution, that of the „discovery of mind".

---

119 WILSON (1978), p. 170.

The comparison of world views demonstrates the primacy of practical reason in decisions concerning theories of totality. Since the self-design of our life and our praxis are themselves a part of the total reality the theory aims for, and since the theory has consequences for our *praxis*, every theory of totality must include itself and the consequent *praxis* resulting from it. The theory must be self-referential. If, within the materialistic evolutionary epic, reflections are made about the theory itself, and if the question is posed whether it serves the praxis of individual and societal human perfection, it becomes evident that it is inferior to the „idealistic“ world view of metaphysics.

To raise question of a theory's application to itself and to human praxis means, since Kant, the recognition of the primacy of practical reason in a theory of totality. According to Kant, the theoretical reason excludes the faculty of the human will to determine its own action, because, in the theoretical reason, the principle of causality holds without exception. But since, according to Kant, this faculty is a „fact of reason“ that cannot be denied, an inclusive systematic theory encompasses theoretical and practical reason and brings theoretical truth and moral goodness into a synthesis.

In his *Critique of Pure Reason* (1781)[120] Kant demonstrates that the theoretical reason is led into antinomies and contradictions when it reflects on the metaphysical questions whether the world has a beginning or is infinite, whether God, an immortal soul and human freedom, the propositions of *metaphysica specialis*, exist or not. According to his theoretical philosophy (i.e. his philosophy of science) we cannot decide these questions by means of categorial understanding. Categorial reasoning, understanding, is led into antinomies if it thinks out to its conclusion any of the metaphysical propositions. Since it is characteristic of understanding to proceed according to the principle of causality and to find a condition for each conditional being, it is led into an infinite regress if understanding applies this principle to the finiteness or infiniteness of the world.

Reason, however, is characterized by the interest in finding the unconditioned at the beginning of the row of conditioned or caused beings. In theoretical reason or philosophy the metaphysical questions of the beginning or infinity of the world, of the existence or non-existence of God, soul, and freedom remain unsolvable. The only possible way to reconcile categorial understanding in its search for causality without gaps and the interest of reason in

120 I. KANT: *Critique of Theoretical Reason*, Riga (Hartknoch) first edition 1781, p. A 426ff.

totality is to ascribe both approaches to reality - that of causal analysis and explanation and that of reflection by the reason about totality and its comprehension by ideas - to two strata of our mental apparatus. Ultimately, this means that the metaphysical questions have to remain undecided in theoretical reason.

In the *Critique of Practical Reason* Kant contends, however, that the propositions of metaphysics, God, freedom, and the immortality of the soul, are postulates of practical reason which for ethical reasons and as conditions conducive to morality we ought to consider real. On the individual ethical level Kant's arguments for the postulates of practical reason are the same as those advanced above as politico-sociobiological arguments forming the conditions for a rational and humane social order. Since the theoretical status of metaphysics is left undecided in Kant's theoretical philosophy but their status in the practical reason well-founded, their over-all philosophical justification is better than is often admitted in the literature on Kant in analytical philosophy. Kant's philosophy is a critique of the theoretical claims of metaphysics as well as it is a possibility theorem for metaphysics and a demonstration of its postulatory reality since it is practical reason which is primary.

In a theory of totality the primacy of practical reason means that the opposition of „true“ and „good“ is not final, but to be overcome: that which is true in a comprehensive sense is also good for the whole and for man. Likewise that which is good for the whole must also be true. An ontology of the selfish gene is not a theory of totality, because it is unable to represent moral *praxis* and the self-understanding of man as a spiritual and free being. We are therefore not compelled by a logic of the theory to adopt the world view of the evolutionary epic and the selfish gene, although it does not forward human sociocultural development. For a theory of the whole, there is no criterium of truth outside of the whole and no criterium other than the whole. The whole, including the practice resulting from the theory, is the criterium of truth.

As a theory of totality, a genuine sociobiological synthesis would have to adequately represent our self-experience as acting and conscious beings and the experience of the social and human world we live in, and it would have to promote or at least not hinder right praxis. In its monistic world-view, the evolutionistic sociobiology does not fulfill this holistic criterium. Its goal of a comprehensive synthesis of natural and social science is nevertheless a speculative goal that must be taken serious and also be pursued by philosophy. The unification of the results from empirical sociobiology with the insights of social philosophy and metaphysics to form a theory of totality is a philosophical task yet to be fulfilled.

References

ALCHIAN, A. A.: „Uncertainty, Evolution, and Economic Theory", *Journal of Political Economy*, 58 (1950), pp. 211-221.

ALEXANDER, R. D.: „The Evolution of Social Behavior", *Annual Review of Ecology and Systematics*, 5 (1974), pp. 325-383.

ALLEN, E. et al. (SOCIOBIOLOGY STUDY GROUP OF SCIENCE FOR THE PEOPLE): „Sociobiology - another biological determinism", *BioScience*, 26 (1976), pp. 182, 184-186.

ARISTOTLE: *De Anima*, J. A. SMITH (tr.), *The Works of Aristotle translated into English*, W. D. ROSS (ed.), Vol. 3, London (Oxford University Press) 1931.

ARISTOTLE: *Nicomachean Ethics*, J. I. BEARE (tr.), *The Works of Aristotle translated into English*, W. D. ROSS (ed.), Vol. 9, London (Oxford University Press) 1925.

BARASH, D. P.: *Sociobiology and Behavior*, New York (Elsevier North Holland) 1977.

BECKER, G.: „Altruism, Egoism, and Genetic Fitness: Economics and Sociobiology", *Journal of Economic Literature*, 14 (1976), pp. 817-826.

BELLAH, R.: *Beyond Belief. Essays on Religion in a Post-Traditional World*, New York (Harper & Row) 1970.

BOCK, K.: *Human Nature and History. A Response to Sociobiology*, New York (Columbia University Press) 1980.

BOULDING, K. E.: „Economics as a Moral Science", *American Economic Review*, 59 (1969).

BOULDING, K. E.: *Evolutionary Economics*, Beverly Hills (Sage) 1981.

BUCHANAN, J. M.: *The Limits of Liberty*, Chicago (University of Chicago Press) 1975.

BUCHANAN, J. M.: „The Related but Distinct ‚Sciences' of Economics and of Political Economy", *British Journal of Social Psychology*, 21 (1982), pp. 175-183.

BUCKLEY, W.: „Sociostructural Systems and the Challenge of Sociobiology", in: H. HAKEN (ed.): *Synergetics. A Workshop*, Berlin (Springer) 1977.

CHAGNON, N. A. / W. IRONS (eds.): *Evolutionary Biology and Human Social Behavior. An Anthropological Perspective*, North Scituate, Mass. (Duxbury) 1979.

CHERRY, R.: „Biology, Sociology, and Economics - An Historical Analysis", *Review of Social Economy*, 38 (1980), pp. 141-154.

CONRAD-MARTIUS, H.: *Utopien der Menschenzüchtung. Der Sozialdarwinismus und seine Folgen*, München (Kösel) 1955.

DAWKINS, R.: *The Selfish Gene*, Oxford (Oxford University Press) 1976, 2nd ed. 1978.

DAWKINS, R.: *The Extended Phenotype*, San Francisco (Freeman) 1982.

DESCARTES, R.: *Discourse on Method*, D. CRESS (ed.), Indianapolis (Hackett) 1980. French original: *Discours de la méthode*, Leyden 1637.

DÖBERT, R.: „Zur Logik des Übergangs von archaischen zu hochkulturellen Religionssystemen", in: KLAUS EDER (ed.): *Seminar: Die Entstehung von Klassengesellschaften*, Frankfurt (Suhrkamp) 1973.

EIGEN, M. / WINKLER, R.: *Laws of the Game: How the Principles of Nature Govern Chance*, New York (Knopf) 1981. Original: EIGEN, M. / WINKLER, R.: *Das Spiel. Naturgesetze steuern den Zufall*, München (Piper) 1975.

FELLMANN, F.: „Darwins Metaphern", *Archiv für Begriffsgeschichte*, 21 (1977), pp. 285-297.

FISHER, R. A.: *The Genetical Theory of Natural Selection*, Oxford (Oxford University Press) 1930.

FRIEDMAN, M.: *Capitalism and Freedom*, Chicago (University of Chicago Press) 1981.

GHISELIN, M.: *The Economy of Nature and the Evolution of Sex*, Berkeley (University of California Press) 1974.

GOULD, S.: „Evolutionäre Flexibilität und menschliches Bewußtsein", in: P. KOSLOWSKI / PH. KREUZER / R. LÖW (eds.): *Evolution und Freiheit*, Stuttgart (S. Hirzel) 1984, pp. 24-35.

GOWDY, J. M.: „Bioeconomics: A Comment", *Review of Social Economy*, 38 (1980), pp. 95-96.

GRAY, J. P. / WOLFE, L. D.: „Sociobiology and Creationism. Two Ethno-Sociologies of American Culture", *American Anthropologist*, 84 (1982), pp. 580-594.

GREGORY, M. S. / SILVERS, A. / SUTCH, D. (eds.): *Sociobiology and Human Nature*, San Francisco (Jossey-Bass) 1978.

GUEROULT, M.: „Note sur le principe de la moindre action chez Maupertuis", in: M. GUEROULT: *Leibniz. Dynamique et Métaphysique*, Paris (Aubier-Montaigne) 1967.

HAMILTON, W. D.: „The Genetical Theory of Social Behaviour (I and II)", *Journal of Theoretical Biology*, 7 (1964), pp. 1-16, 17-32.

HAYEK, F. A. V.: *Die Theorie komplexer Phänomene*, Tübingen (J.C.B. Mohr [Paul Siebeck]) 1972.

HAYEK, F. A. V.: *Mißbrauch und Verfall der Vernunft* (1959), München (Philosophia) 1974.

HAYEK, F. A. V.: *The Counter-Revolution of Science*, Indianapolis (Liberty Classics) 1980.

HEGEL, G.W.F.: *Science of Logic*, A. MILLER (tr.), Atlantic Highlands (Humanities Press) 1976.

REFERENCES

HEGEL, G.W.F.: *Philosophy of Right*, T. M. KNOX (tr.), London (Oxford University Press) 1942.

HENRICH, D.: „Absoluter Geist und Logik des Endlichen", *Hegel-Studien*, Beiheft 20, Bonn 1980, pp. 103-118.

HETTLAGE, R.: „Variationen des Darwinismus in der Soziologie", in: D. HENRICH (eds.): *Evolutionstheorie und ihre Evolution*, Regensburg (Schriftenreihe der Universität Regensburg Bd. 7) 1982, pp. 109-125.

HIRSHLEIFER, J. (1978a): „Competition, Cooperation, and Conflict in Economics and Biology", *American Economic Review*, 68 (1978).

HIRSHLEIFER, J. (1978b): „Natural Economy Versus Political Economy", *Journal of Social and Biological Structures*, 1 (1978), pp. 319-337.

HIRSHLEIFER, J.: „Evolution, spontane Ordnung und Marktwirtschaft", in: P. KOSLOWSKI / PH. KREUZER / R. LÖW (eds.): *Evolution und Freiheit*, Stuttgart (S. Hirzel) 1984, pp. 77-92.

HOPPMANN, E.: „Gleichgewicht und Evolution. Voraussetzungen und Erkenntniswert der volkswirtschaftlichen Totalanalyse", in: *Festschrift für Erich Carell*, Baden Baden (Nomos) 1980.

JÜNGER, E.: *Jahre der Okkupation*, Stuttgart (Klett) 1958.

KANT, I.: *Critique of Practical Reason* (1788), L. W. Beck (tr.), New York (Bobbs) 1956.

KANT, I.: *Critique of Pure Reason* (1781), N. K. Smith (ed.), London/New York (St. Martin's Press) 1969.

KNIGHT, F. H.: „Ethics and the Economic Interpretation", in: F. H. KNIGHT: *The Ethics of Competition and Other Essays* (1935), Reprint Freeport (Books for Libraries Press) 1969.

KOCH, H.: *Der Sozialdarwinismus*, München (C. H. Beck) 1973.

KOHLBERG, L.: *Essays on Moral Development*, San Francisco (Harper & Row) 1981.

KOSLOWSKI, P. (1982): *Gesellschaft und Staat. Ein unvermeidlicher Dualismus*, Stuttgart (Klett-Cotta) 1982.

KOSLOWSKI, P. (1983a): „Markt- und Demokratieversagen? Grenzen individualistischer gesellschaftlicher Entscheidungssysteme", *Politische Vierteljahresschrift*, 24 (1983), pp. 166-187.

KOSLOWSKI, P. (1983b): „Mechanistische und organistische Analogien in der Wirtschaftswissenschaft - eine verfehlte Alternative", *Kyklos*, 36 (1983), pp. 308-312.

KOSLOWSKI, P., PH. KREUZER, R. LÖW (eds.): *Evolution und Freibeit*, Stuttgart (S. Hirzel) 1984.

KOSLOWSKI, P. (1985): „Economy Principle, Maximizing, and the Co-ordination of Individuals in Economics and Philosophy", in: P. KOSLOWSKI (ed.): *Economics and Philosophy*, Tübingen (J.C.B. Mohr [P. Siebeck]) 1985, pp. 39-67.

KRINGS, H.: *Meditation des Denkens*, München (Kösel) 1956.

KRINGS, H.: „Wissen und Freiheit" (1966), in: H. KRINGS: *System und Freibeit*, Freiburg (Alber) 1980, pp. 133-160.

KRINGS, H.: „Evolution und Revolution - Zwei Interpretamente der modernen Welt", in: R. LÖW/ P. KOSLOWSKI/ PH. KREUZER (eds.): *Fortschritt ohne Maß? Eine Ortsbestimmung der wissenschaftlich-technischen Zivilisation*, München (Piper) 1981, pp. 29-47.

KRINGS, H.: „Sokrates überlebt. Zum Verhältnis von Evolution und Geschichte", in: P. KOSLOWSKI / PH. KREUZER / R. LÖW (eds.): *Evolution und Freibeit*, Stuttgart (S. Hirzel) 1984, pp. 162-176.

LEHMANN, M.: „Jurisprudenz, Ökonomie, Soziobiologie - Humanwissenschaften. Über die Notwendigkeit und Möglichkeit einer interdisziplinären Grundlagenforschung", *Betriebs-Berater*, Heft 33 (1982), pp. 1997-2004.

LEIBNIZ, *Letter to Clarke*, quoted in M. GUEROULT: *Leibniz. Dynamique et Métaphysique*, Paris (Aubier-Montaigne) 1967.

LEVI-STRAUSS, C.: *Structural Anthropology*, Vol. 1, New York (Basic) 1963, Vol. 2, Chicago (University of Chicago Press) 1976.

LEWONTIN, R. C.: „Adaptation", *Scientific American*, 239 (1978), pp. 157-169.

LEWONTIN, R. C.: „Fitness, Survival, and Optimality", in: DAVID J. HORN / G. R. STAIRS / R. D. MITCHELL (eds.): *Analysis of Ecological Systems*, Columbus (Ohio State University Press) 1979.

LÖW, R.: *Philosophie des Lebendigen. Der Begriff des Organischen bei Kant*, Frankfurt (Suhrkamp) 1980.

LORENZ, K.: *Behind the Mirror*, San Diego/New York (Harcourt Brace Jovanovich) 1977. Original: LORENZ, K.: *Die Rückseite des Spiegels. Versuch einer Naturgeschichte menschlichen Erkennens*, München (Piper) 1973.

LUCE, R. D. / H. RAIFFA: *Games and Decisions*, New York (John Wiley) 1957.

LUHMANN, N.: „Evolution und Geschichte", in: N. LUHMANN: *Soziologische Aufklärung* 2, Opladen (Westdeutscher Verlag) 1975.

LUMSDEN, CH. / E. O. WILSON: *Genes, Mind, and Culture. The Coevolutionary Process*, Cambridge Mass. (Harvard University Press) 1981.

MARKIS, D.: Artikel „Rekonstruktion", in: E. BRAUN / H. RADERMACHER: *Wissenschaftstheoretisches Lexikon*, Graz (Styria) 1978.

MARKL, H.: *Aggression und Altruismus. Coevolution der Gegensätze im Sozialverhalten der Tiere*, Konstanz (Konstanzer Universitätsreden Bd. 46) 1976.

MARKL, H. (1980a) (ed.): *The Evolution of Social Behavior. Hypotheses and Empirical Tests*, Weinheim (Verlag Chemie) 1980.

MARKL, H. (1980b): „Ökologische Grenzen und Evolutionsstrategie Forschung", *Forschung-Mitteilungen der DFG*, 3 (1980).

MARKL, H.: „The Power of Reduction and the Limits of Compressibility", *The Behavioral and Brain Sciences*, 5 (1982), pp. 18 f.

MARSHAK, J.: „Economics of Acting, Thinking and Surviving", in: J. MARSHAK: *Economic Information, Decision and Prediction*, Vol. I, Dordrecht (Reidel) 1974.

REFERENCES

MARTEN, H.-G.: *Sozialbiologismus. Biologische Grundpositionen der politischen Ideengeschichte,* Frankfurt (Campus) 1982.

MASTERS, R. D.: „Sociobiology: science or myth?", *Journal of Social and Biological Structures,* 2 (1979), pp. 245-252.

MAUPERTUIS, P. L. MOREAU DE: *Essai de Cosmologie* (1759), in: P. L. MOREAU DE MAUPERTUIS: *Oeuvres,* Lyon 1768, Reprint Hildesheim (Olms) 1974.

MAYNARD SMITH, J.: „Evolution and the Theory of Games", *American Scientist,* 64 (1976).

MAYR, E.: *Animal Species and Evolution,* Cambridge (Harvard University Press) 1963.

MCCAIN, R. A.: „Critical Reflections on Sociobiology", *Review of Social Economy,* 38 (1980), pp. 123-139.

MERTON, R. K.: *Social Theory and Social Structure,* Glencoe (Free Press) 1957.

NAGEL, T.: „Ethics as an Autonomous Theoretical Subject", in: G. S. STENT: *Morality as a Biological Phenomenon. The Presuppositions of Sociobiological Research,* Berkeley (University of California Press) 1980.

OSTER, G. / E. O. WILSON: *Caste and Ecology in the Social Insects,* Princeton (Princeton University Press) 1978, especially chapter 8: „A Critique of Optimization Theory in Evolutionary Biology".

PETERS, H. M.: „Soziomorphe Modelle in der Biologie", *Ratio,* 3 (1960), pp. 22-37.

PETERS, H. M.: „Historische, soziologische und erkenntniskritische Aspekte der Lehre Darwins", in: HANS-GEORG GADAMER / P. VOGLER: *Neue Anthropologie,* Bd. I, Stuttgart (Thieme/dtv) 1972, pp. 326-352.

PLATO: *The Republic (Politeia),* J. ADAM (ed.), Cambridge (Cambridge University Press) 1965.

PORKSEN, U.: „Die Metaphorik Darwins und Überlegungen zu ihrer möglichen Wirkung", in: *Wissenschaftskolleg Jahrbuch 1981/82,* hrsg. von PETER WAPNEWSKI, Berlin (Quadriga) 1983, pp. 256-280.

RAMSEY, F. P.: „Truth and Probabilities" (1926), in: F. P. RAMSEY: *The Foundations of Mathematics and Other Logical Essays,* London (Routledge & Kegan Paul) 1978.

RAPOPORT, A. / A. M. CHAMMAH: *Prisoner's Dilemma. A Study in Conflict and Cooperation,* Ann Arbor (University of Michigan Press) 1956.

RAPPORT, D. J. / J. E. TURNER: „Economic Models in Ecology", *Science,* 195 (1977), pp. 367-373.

ROSENBERG, A.: *Sociobiology and the Preemption of Social Science,* Baltimore (Johns Hopkins University Press) 1980.

SAHLINS, M.: *The Use and Abuse of Biology. An Anthropological Critique of Sociobiology,* Ann Arbor (University of Michigan Press) 1976.

SAMUELSON, P. A.: „Maximizing and Biology", *Economic Inquiry,* 16 (1978), pp. 171-183.

SIMON, H. A.: „Rationality as Process and as Product of Thought", *American Economic Review*, 68 (1978), pp. 1-15.

SINGER, P.: *The Expanding Circle. Ethics and Sociobiology*, New York (Farrar, Straus, Giroux) 1981.

SINGER, P.: „Ethics and Sociobiology", *Philosophy and Public Affairs*, 11 (1982), pp. 40-64.

SPAEMANN, R. (1977a): „Die Aktualität des Naturrechts",

SPAEMANN, R. (1977b): „Praktische Gewißheit. Descartes' provisorische Moral", both articles in: R. SPAEMANN: *Kritik der politischen Utopie*, Stuttgart (Klett) 1977.

SPAEMANN, R.: „Über die Unmöglichkeit einer universalteleologischen Ethik", *Philosophisches Jahrbuch*, 88 (1981), pp. 70-89.

SPAEMANN, R. / R. LÖW: *Die Frage Wozu? Geschichte und Wiederentdeckung des teleologischen Denkens*, München (Piper) 1981, 2nd ed. 1985.

THOMAS, W. I.: *Social Behaviour and Personality: Contributions of W. I. Thomas to Theory and Social Research*, E. H. VOLKART (ed.), Westport (Gruwood) 1981. First edition 1928.

TOPITSCH, E.: „Das Verhältnis zwischen Sozial- und Naturwissenschaften. Eine methodologisch-ideologiekritische Untersuchung", *Dialectica*, 16 (1962), pp. 211-231.

TRIVERS, R. L.: „The Evolution of Reciprocal Altruism", *Quarterly Review of Biology*, 46 (1971), pp. 35-57.

TULLOCK, G.: „An Application of Economics in Biology", in: *Toward Liberty. Essays in Honor of Ludwig von Mises on the Occasion of his 90th Birthday*, Menlo Park (Institute for Humane Studies) 1971, Vol. II, pp. 375-391.

TULLOCK, G.: „Economics and Sociobiology. A Comment", *Journal of Economic Literature*, 15 (1977), pp. 502-506.

VANBERG, V.: „Libertarian Evolutionism and Contractarian Constitutionalism", in: SVETOZAR PEJOVICH (ed.): *Philosophical and Economic Foundations of Capitalism*, Lexington (D. C. Heath) 1983, pp. 71-88.

VOSSENKUHL, W.: „Die Unableitbarkeit der Moral aus der Evolution", in: P. KOSLOWSKI / PH. KREUZER / R. LÖW (eds.): *Die Verführung durch das Machbare. Ethische Konfliite in der modernen Medizin und Biologie*, Stuttgart (S. Hirzel) 1983, pp. 141-154.

WHEELER, H.: „Human Sociobiology. An Exploratory Essay", *Journal of Social and Biological Structures*, 1 (1978), pp. 307-318.

WICKLER, W.: „Hat die Ethik einen evolutionären Ursprung?", in: P. KOSLOWSKI / PH. KREUZER / R. LÖW (eds.): *Die Verführung durch das Machbare. Ethische Konflikte in der modernen Medizin und Biologie*, Stuttgart (S. Hirzel) 1983, pp. 125-140.

WICKLER, W. / U. SEIBT: *Das Prinzip Eigennutz* (1977), München (dtv) 1981.

WILSON, E. O.: *Sociobiology. The New Synthesis*, Cambridge (Harvard University Press) 1975.

REFERENCES

WILSON, E. O.: „Academic Vigilantism and the Political Significance of Sociobiology", *BioScience*, 26 (1976), pp. 183, 187-190.

WILSON, E. O.: *On Human Nature*, Cambridge (Harvard University Press) 1978.

Index of Names

Page numbers in italics refer to quotations in footnotes or references

INDEX OF NAMES

GPSR Compliance

The European Union's (EU) General Product Safety Regulation (GPSR) is a set of rules that requires consumer products to be safe and our obligations to ensure this.

If you have any concerns about our products, you can contact us on ProductSafety@springernature.com

In case Publisher is established outside the EU, the EU authorized representative is:

Springer Nature Customer Service Center GmbH
Europaplatz 3
69115 Heidelberg, Germany

Batch number: 10844675

Printed by Printforce, the Netherlands